Steve Smale

The Mathematics of Time

Essays on Dynamical Systems, Economic Processes, and Related Topics

Springer-Verlag
New York Heidelberg Berlin

Steve Smale
Department of Mathematics
University of California at Berkeley
Berkeley, CA 94720
USA

¥ 3 ·ᴅʀ 198ⁱ 275769

515·35 sMA

AMS Classifications: 00A10, 58-01, 90A16

Library of Congress Cataloging in Publication Data

Smale, Stephen, 1930–
   The mathematics of time.

   Includes bibliographical references.
   1. Global analysis (Mathematics)—Addresses, essays,
lectures. 2. Economics, Mathematical—Addresses, essays,
lectures. I. Title.
QA614.S6   514'.74   80-20827

With 26 figures.

Printed in the United States of America.

9 8 7 6 5 4 3 2 1

ISBN   0-387-90519-7 Springer-Verlag New York
ISBN   3-540-90519-7 Springer-Verlag Berlin Heidelberg

Contents

Permissions

Differential Dynamical Systems
Reprinted with permission of the publisher, The American Mathematical Society, from *Bulletin of the American Mathematical Society*, Copyright © 1967, Volume 73, pp. 747–817.

What is Global Analysis?
Reprinted with permission of the publisher, The Mathematical Association of America, from the *American Mathematical Monthly*, Copyright © 1969, Volume 76, #1, pp. 4–9.

Stability and Genericity in Dynamical Systems
Reprinted with permission of the publisher, the Association Nicolas Bourbaki, from *Seminaire Bourbaki*, Copyright © 1970, Volume #374 (Lecture Notes in Mathematics, Volume 1969/70, #180).

Personal Perspectives on Mathematics and Mechanics
Reprinted with permission of the publisher, the University of Chicago Press, from *Statistical Mechanics*: New Concepts, New Problems, New Applications: Proceedings of the Sixth IUPAP Conference on Statistical Mechanics, edited by Rice, Light and Freed, 1972.

Dynamics in General Equilibrium
Reprinted with permission of the publisher, the American Economic Association, from the *American Economic Review*, Copyright © 1976, Volume 66, #2, pp. 284–294.

Some Dynamical Questions in Mathematical Economics
Reprinted with permission of the publisher, Centre National de la Recherche Scientifique, from *Systemes Dynamiques et Modeles Economiques*, Colloque 259, Copyright © 1977.

Review of *Global Variational Analysis: Weierstrass Integrals on a Reimannian Manifold*
Reprinted with permission of the publisher, American Mathematical Society, from *Bulletin of the American Mathematical Society*, Copyright © 1977, Vol. 83, #4, pp. 683–693.

DIFFERENTIABLE DYNAMICAL SYSTEMS[*],[1]

PART I. DIFFEOMORPHISMS

I.1. Introduction to conjugacy problems for diffeomorphisms. This is a survey article on the area of global analysis defined by differentiable dynamical systems or equivalently the action (differentiable) of a Lie group G on a manifold M. An action is a homomorphism $G \to \mathrm{Diff}(M)$ such that the induced map $G \times M \to M$ is differentiable. Here $\mathrm{Diff}(M)$ is the group of all diffeomorphisms of M and a diffeomorphism is a differentiable map with a differentiable inverse. Everything will be discussed here from the C^∞ or C^r point of view. All manifolds maps, etc. will be differentiable (C^r, $1 \leq r \leq \infty$) unless stated otherwise.

In the beginning we will be restricted to the discrete case, $G = Z$. Here Z denotes the integers, Z^+ the positive integers. By taking a generator $f \in \mathrm{Diff}(M)$, this amounts to studying diffeomorphisms on a manifold from the point of view of orbit structure. The orbit of $x \in M$, relative to f, is the subset $\{f^m(x) \mid m \in Z\}$ of M or else the map $Z \to M$ which sends m into $f^m(x)$. The finite orbits are called *periodic orbits* and their points, *periodic points*. Thus $x \in M$ is a periodic point if $f^m(x) = x$ for some $m \in Z^+$. Here m is called a period of x and if $m = 1$, x is a *fixed point*. Our problem is to study the global orbit structure, i.e., all of the orbits on M.

The main motivation for this problem comes from ordinary differential equations, which essentially corresponds to $G = R$, R the reals acting on M. There are two reasons for this leading to the diffeomorphism problem. One is that certain differential equations have cross-sections (see, e.g., [114]) and in this case the qualitative study of the differential equation reduces to the study of an associated diffeomorphism of the cross-section. This is the reason why Poincaré [90] and Birkhoff [19] studied diffeomorphisms of surfaces.

I believe there is a second and more important reason for studying the diffeomorphism problem (besides its great natural beauty). That is, the same phenomena and problems of the qualitative theory of ordinary differential equations are present in their simplest form in the diffeomorphism problem. Having first found theorems in the dif-

* The preparation of this paper was supported by the National Science Foundation under grant GN-530 to the American Mathematical Society and partially supported by NSF grant GP-5798.

feomorphism case, it is usually a secondary task to translate the results back into the differential equations framework.

Assuming M compact, we put on Diff(M) the topology of uniform C^r convergence. We will usually keep M compact because for noncompact M, there are different behaviours at infinity that one could consider. See, for example, [86]. These lead to different problems and we don't wish to get into such questions here.

One of the first things that one observes is the need to exclude degenerate elements of Diff(M). For example, given any nonempty closed set $F \subset M$, there is $f \in$ Diff(M) such that the fixed point set Fix(f) $= F$. For a number of reasons, if F is not discrete we would like to exclude such f. The set of $f \in$ Diff(M) such that Fix(f) is discrete (or finite, since we assume M compact) contains an open dense set of Diff(M). This leads to the notion of generic properties of diffeomorphisms. A *Baire set* of a complete metrizable space is the intersection of a countable number of open dense sets. Then a *generic property* is a property that is true for diffeomorphisms belonging to some Baire set of Diff(M). We will never speak of generic $f \in$ Diff(M) (this is usually taken to mean that f has a lot of generic properties!). Thus "Fix(f) is finite" is a generic property and a little more since open dense is stronger than Baire (see §I.6 for more details and references).

It is important in proceeding to consider formal equivalence relations on Diff(M) which will preserve the orbit structure in some sense. Furthermore associated to each equivalence relation there is a notion of stability. More precisely if the equivalence relation on Diff(M) is called E, $f \in$ Diff(M) is called *E-stable* if there is a neighborhood $N(f)$ of f in Diff(M) such that if $f' \in N(f)$ (or f' approximates f sufficiently), then f and f' are in the same E equivalence class.

It would give a reasonable picture (see [111], [112]) to have a dense open set $U \subset$ Diff(M) such that our equivalence classes could be distinguished by numerical and algebraic invariants. This is, in fact, our goal.[2] If this is to be the case, the desired equivalence E on Diff(M) should have the property that the E-stable diffeomorphisms are dense in Diff(M). With this background we look at some particular equivalence relations.

The notion of conjugacy first comes to mind. Say f, $f' \in$ Diff(M) are differentiably (or topologically) *conjugate* if there is a diffeomorphism (or homeomorphism) $h: M \rightarrow M$ such that $hf = f'h$. Differentiable conjugacy is too fine in view of the above considerations. This is due to the fact that the eigenvalues of the derivative at a fixed point are differentiable conjugacy invariants. The notion of stability

2

associated to topological conjugacy is called structural stability, and for some time it was thought that structurally stable diffeomorphisms might be dense in Diff(M). This turned out to be false [116]. Thus by our earlier consideration we should relax our relation on Diff(M) of topological conjugacy. Before doing this we introduce some basic ideas about G. D. Birkhoff's nonwandering points [15].

If $f \in$ Diff(M), $x \in M$ is called a *wandering point* when there is a neighborhood U of x such that $\bigcup_{|m|>0} f^m(U) \cap U = \emptyset$. The wandering points clearly form an invariant open subset of M. A point will be called *nonwandering* if it is not a wandering point. These nonwandering points are those with the mildest possible form of recurrence. They form a closed invariant set which we will always refer to as $\Omega = \Omega(f)$.[3]

We propose now the equivalence "topological conjugacy on Ω." That is f, $f' \in$ Diff(M) are *topologically conjugate* on Ω if there is a homeomorphism $h: \Omega(f) \to \Omega(f')$ such that $hf = f'h$. The corresponding stability will be called simply Ω-*stability*. So $f \in$ Diff(M) will be called Ω-stable if sufficiently good approximations f' are topologically conjugate on Ω.

In general one can speak of topological conjugacy for homeomorphisms and even two homeomorphisms of different topological spaces, $f: X \to X$, $f': X' \to X'$. Then the conjugacy h is a homeomorphism $h: X \to X'$.

We end §I.1 by giving some notations and conventions we follow.

Anytime the topology on Diff(M) is involved M will be assumed compact.

Simply connected X means $\Pi_1(X)$ and $\Pi_0(X)$ are trivial. We suppose that our manifolds are always connected.

Dim M means the dimension of M.

The tangent bundle of a manifold will be denoted by $T(M)$, the tangent space at $x \in M$ by $T_x(M)$. The *derivative* of $f: M \to M$ will be denoted by Df and considered as a bundle map $Df: T(M) \to T(M)$. At a point $x \in M$, it becomes $Df(x): T_x(M) \to T_{f(x)}(M)$. An *immersion* is a differentiable map such that the derivative at each point is injective.

A closed invariant set Λ of $f \in$ Diff(M) will be called indecomposable if Λ cannot be written $\Lambda = \Lambda_1 \cup \Lambda_2$, Λ_1, Λ_2 nonempty disjoint closed invariant subsets.

Finally if λ is an eigenvalue of a linear transformation $u: V \to V$, we will define its eigenspace $E_\lambda = \{x \in V \mid (u - \lambda I)^m(x) = 0$, some $m \in Z^+\}$. Then λ will be counted with multiplicity dim E_λ.

Two earlier surveys on this subject are [85] and [112].

Part I is the heart of the paper, including a number of new ideas, and is devoted to problems spoken of in this section. Part II briefly extends the results to the ordinary differential equation case $(G=R)$ and Part III discusses other aspects of the differential equation problem. Part IV is devoted to possibilities for more general Lie groups G.

I would like to acknowledge here many very helpful discussions with other mathematicians. This includes especially D. Anosov, A. Borel, A. Haefliger, M. Hirsch, N. Kopell, I. Kupka, J. Moser, R. Narasimhan, J. Palis, M. Peixoto, C. Pugh, M. Shub and R. Thom.

I.2. The simplest examples. This section is devoted to giving a description of a class of Ω-stable diffeomorphisms which are the simplest as far as the orbit structure goes. To develop or even define these diffeomorphisms, we will need the basic idea of a stable manifold.

A linear automorphism u of a (say real) finite dimensional vector space V, $u: V \rightarrow V$ will be called *hyperbolic* if its eigenvalues λ_i satisfy $|\lambda_i| \neq 1$ all i. We emphasize that complex eigenvalues are permitted. The automorphism u will be called *contracting* if $|\lambda_i| < 1$ for all i, *expanding* if $|\lambda_i| > 1$ for all i, and of *saddle type* otherwise. Thus the inverse of an expanding automorphism is a contracting automorphism and vice versa.

Observe that for hyperbolic $u: V \rightarrow V$ we have a canonical, invariant (under u) splitting of V, $V = V^s + V^u$ (direct sum) where V^s is the eigenspace of u corresponding to eigenvalues less than 1 in absolute value and V^u the eigenspace of the remaining eigenvalues. Thus u restricted to V^s is contracting and u restricted to V^u is expanding. This gives rise to the following familiar picture for such u.

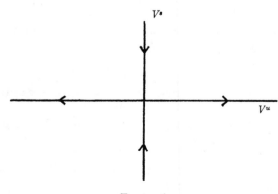

FIGURE 1

Note that the hyperbolic elements of the general linear group $GL(V)$ are open and dense.

Now suppose $f: M \rightarrow M$ is a diffeomorphism with a fixed point $p \in M$ (a local diffeomorphism $f: U \rightarrow M$, U an open subset of M, $p \in U$, $f(p) = p$ would be sufficient for some of the following discussion). The derivative of f at p, $Df(p)$, may be considered to be a linear automorphism of the tangent space of M at p, i.e., $Df(p): T_p(M) \rightarrow T_p(M)$. We will say that p is a *hyperbolic fixed point* of f, or simply a hyperbolic fixed point, if $Df(p)$ is hyperbolic in the sense of the previous paragraphs.

We will call a periodic point p of period $m \in Z^+$ of $f: M \rightarrow M$ hyperbolic if it is a hyperbolic fixed point of f^m. Similarly, p is a *contracting* or *expanding* periodic point if $Df^m(p)$ is a contracting (or expanding) linear automorphism.

A (global) *contraction* of a differentiable manifold V is a diffeomorphism $g: V \rightarrow V$ which is topologically conjugate to a linear contraction (i.e., a linear contracting automorphism) $u: V' \rightarrow V'$. Of course a contraction will have a unique fixed point.

For hyperbolic fixed points we have stable manifolds defined according to the following theorem.

(2.1) *Stable Manifold Theorem*. Suppose $p \in M$ is a hyperbolic fixed point of a diffeomorphism $f: M \rightarrow M$ with $T_p(M) = V^s + V^u$ the corresponding decomposition under $Df(p)$. Then there exists a contraction $g: W^s(p) \rightarrow W^s(p)$ with fixed point p_0 and an injective equivariant immersion $J: W^s(p) \rightarrow M$ such that $J(p_0) = p$ and $DJ(p_0): T_{p_0}(W^s(p)) \rightarrow T_p(M)$ is an isomorphism onto $V^s \subset T_p(M)$. Furthermore the image $J(W^s(p))$ may be characterized as the set of $x \in M$ with the property $f^m(x) \rightarrow p$ as $m \rightarrow \infty$.

Equivariance here means simply that $Jg = fJ$. Note that the derivative condition implies that the dimensions of V^s and $W^s(p)$ are the same.

The image of J is invariant under f, and frequently we will identify points under J so that $W^s(p) \subset M$. In general, J will not be a homeomorphism onto its image (see the toral example of §I.3), so that the original $W^s(p)$ and $W^s(p)$ is a subset of M have different topologies and this is the only way they differ. Both are called the *stable manifold* of f at p. When it is important to specify the topology, we will say *intrinsic* for the original topology and the *manifold* topology for the other topology on $W^s(p)$.

For analytic diffeomorphisms of two dimensional manifolds, this theorem was known to Poincaré [90] and used by Birkhoff [16].

The proof of (2.1) starts by showing the existence of a "local stable manifold," $W^s_{loc}(p)$. This is due to Perron [88]. He uses iteration methods in a function space to solve a functional equation for J in a neighborhood of p. Further references to versions of this theorem are [2], [24], [39], [120] (most often these papers concern themselves with the differential equations analogue, so one has to make a translation of the results). The global theorem, (2.1), follows easily from the local theorem by so to speak "topological continuation." One takes for $W^s(p)$ the subset $\cup_{m \in Z^+} f^{-m} W^s_{loc}(p)$ of M. See [114] for more details.

For a hyperbolic fixed point p of a diffeomorphism $f: M \rightarrow M$, the *unstable manifold* $W^u(p)$ is defined as the stable manifold of f^{-1} at p. Thus $W^u(p)$ passes through p and is tangent to V^u in the notation of (2.1).

For a periodic point q of $f \in \text{Diff}(M)$, $f^m(q) = q$, $m \in Z^+$, one defines the *stable* and *unstable manifolds*, $W^s(p)$, $W^u(q)$ as the stable and unstable manifolds for q as a fixed point of f^m.

Although each $W^s(p)$ is a 1-1 immersion, there is no reason why $W^s(p)$ and $W^u(q)$ cannot intersect each other. In fact as the toral example of §I.3 shows, it may happen that $W^s(p)$ intersects $W^u(p)$ (this is called a homoclinic point; see §I.5).

We now are in a position to describe the examples, or the class of examples, we mentioned earlier. As a prototype it is worthwhile to keep in mind the diffeomorphism $g_0: S^2 \rightarrow S^2$ of the 2-sphere which can be described complex analytically on the Riemann sphere by $z \rightarrow 2z$. The two fixed points are 0 which is expanding and ∞, contracting. Then $W^u(0) = S^2 - \infty$, $W^s(0) = 0$, $W^u(\infty) = \infty$, $W^s(\infty) = S^2 - 0$. It is easily checked that g_0 is structurally stable. Of course one may construct a similar example on S^n with two fixed points.

More generally we will consider $f \in \text{Diff}(M)$, M compact, which satisfies the following three conditions:

(2.2) (1) Ω, the nonwandering set, is finite.

(2) The periodic points of f are hyperbolic.

(3) (Transversal intersection condition) For each p, $q \in \Omega$, $W^s(p)$ and $W^u(q)$ have transversal intersection.

It follows from (1) that Ω consists of periodic points and (2) that $W^s(p)$, $W^u(q)$ are defined for p, $q \in \Omega$. The last condition means that whenever $x \in W^s(p) \cap W^u(q)$, then $T_x(W^s(p))$ and $T_x(W^u(q))$ span $T_x(M)$.

It is trivial to check that the above $g_0: S^2 \rightarrow S^2$ satisfies (1)–(3).

Furthermore, consider for the moment, diffeomorphisms of the circle S^1 satisfying (2.2). In this case (2.2)-(3) is vacuously satisfied, and it is easily checked directly that these diffeomorphisms are open. By perturbing an arbitrary $f \in \text{Diff}(S^1)$ so that its rotation number [24] becomes rational and a further approximation to obtain (2.2)-(1) we obtain the fact that these diffeomorphisms are open and dense in $\text{Diff}(S^1)$ (Peixoto's theorem [84]). As one goes around the circle, the expanding and contracting periodic points alternate. The structural stability in the case is easy to check [84].

If $A \subset B$, clos A denotes the closure of A in B.

(2.3) THEOREM [109].[4] Suppose $f: M \to M$ satisfies (2.2). Then (a) for each $p \in \Omega$, $W^s(p)$ is imbedded in M and $M = \bigcup_{p \in \Omega} W^s(p)$ (disjoint union of course).

(b) clos $W^s(p)$ is the union of $W^s(q)$, for q in some subset of Ω. If we write $\gamma \leq \gamma'$ for periodic orbits γ, γ' whenever $\bigcup_{p \in \gamma} W^s(p) \subset \text{clos} \bigcup_{q \in \gamma'} W^s(q)$, then $\leq$ is a partial ordering. If $\gamma \leq \gamma'$ and $p \in \gamma$, $q \in \gamma'$, then dim $W^s(p) \leq$ dim $W^s(q)$.

(c) One has the following Morse inequalities:

$$M_0 > B_0,$$

$$M_1 - M_0 \geqq B_1 - B_0,$$

$$\vdots$$

$$\sum_{i=0}^{\dim M} (-1)^i M_i = \sum_{i=0}^{\dim M} (-1)^i B_i.$$

Here B_i is the ith betti number of M and M_j is the number of periodic points p such that dim $W^s(p) = j$.

The essence of the proof of (2.3) is in a more general context in §I.8.

Using (2.3) (b), one may "represent" a diffeomorphism satisfying (2.2) by a diagram where the vertices of the diagram correspond to periodic orbits and oriented segments are placed between orbits γ and γ' when $\gamma \leq \gamma'$ but there is no other γ'' such that $\gamma \leq \gamma'' \leq \gamma'$.

A *labeled diagram* is a diagram with the following additional data attached to each vertex γ. The additional data is the germ of the topological conjugacy class of f^m at $x \in \gamma$ where m is the least period of γ. This germ is described precisely by the dimensions of $W^s(x)$, $W^u(x)$ and whether $f: W^u(x) \to W^u(x)$, $f: W^s(x) \to W^s(x)$ are orientation preserving or reversing (this is a consequence of the theorem of Hartman [39] and Grobman (see [74] which says that locally a diffeomorphism at a hyperbolic fixed point is topologically equivalent to its derivative at that point).

(2.4) PROBLEM[5] (a) Exactly what (abstract) labeled diagrams occur as diagrams of diffeomorphisms satisfying (2.2)?

(b) Given compact M exactly what (abstract) labeled diagrams occur as the labeled diagrams of diffeomorphisms of M satisfying (2.2)?

Note that (2.3) (c) may be viewed as a restriction on the kind of diagrams that can occur.

Figure 2 below gives the phase portrait or orbit structure of an example of a diffeomorphism of the 2-sphere satisfying (2.2).

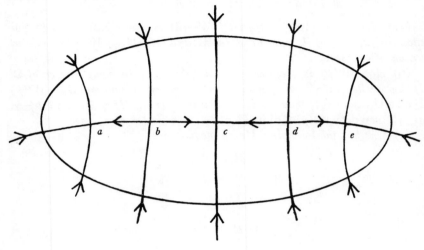

FIGURE 2

Here the main disk is to be contracting into itself with one expanding fixed point p outside. Inside the disk are five fixed points a, c, e all contracting and b, d of saddle type. The diagram for this diffeomorphism is given by

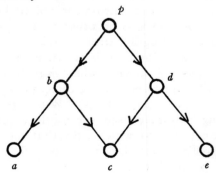

FIGURE 3

Among other interesting results on this subject, Jacob Palis [82] shows that diffeomorphisms satisfying (2.2) form an open set in $\text{Diff}(M)$. He also shows that the diagram of the perturbation of f is naturally "isomorphic" to the diagram of f.

Even though the above facts give something of a "phase portrait" (in the terminology of [57]), a number of problems on this subject still remain. For example

(2.5) PROBLEM[6] [109], [111]. Are these diffeomorphisms (of 2.2) structurally stable? J. Palis has given an affirmative answer in dimension 2.

(2.6)[7] What homotopy classes of continuous maps (homotopy equivalences) admit diffeomorphisms of (2.2) type? A necessary condition which follows from the Lefschetz trace formula is that $|\Lambda(f^m)|$ $<C$, where Λ is the Lefschetz number and C is a constant independent of m.

The gradient-like diffeomorphisms are a special class of diffeomorphisms satisfying (2.2), the most transparent and easily understood. More precisely a *gradient-like* diffeomorphism is one which satisfies (2.2) and has the additional property that if $W^s(p) \leqq W^s(q)$, then dim $W^s(p)$ is actually less than dim $W^s(q)$. For example the diffeomorphism of diagram 2 is gradient-like.

More generally every gradient flow with mild transversality and nondegeneracy conditions (see [110]) generates a gradient-like diffeomorphism. This construction guarantees the existence of gradient-like diffeomorphisms (satisfying (2.2)) on every compact manifold. In this way the above Morse inequalities (2.3) (c) include the usual ones. Even for these diffeomorphisms, structural stability is not yet proved.

For a 2-dimensional diffeomorphism satisfying (2.2) a heteroclinic point is a point $x \in W^u(p) \cap W^s(q)$ where dim $W^s(q) = $ dim $W^s(p)$ $=1$, so that at x, $T_x(W^u(p))$ and $T_x(W^s(q))$ intersect in just one point in $T_x(M)$. Clearly a diffeomorphism possessing a heteroclinic point is not gradient-like. The orbit of a heteroclinic point consists of other heteroclinic points.

The interested reader will be able to check that the existence of the heteroclinic point x above forces $W^s(q)$ to oscillate strongly as it gets close to p and $W^s(p)$. The boundary of $W^s(q)$ contains $W^s(p)$. The picture looks something like Figure 4.

To obtain a global example one may modify the diffeomorphism of the 2-sphere of Figure 2. The result will be something like Figure 5.

9

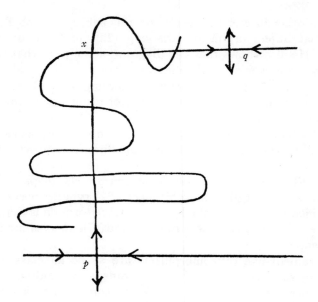

FIGURE 4

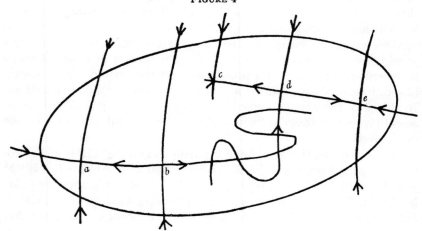

FIGURE 5

Its diagram is given in Figure 6.

We discuss relaxing or dropping some of the conditions (1), (2), (3) of (2.2). The rest of Part I is concerned with weakening (1), so we consider now (2) and (3). It seems to us that dropping (2) or even modifying (2) significantly would take one far from the picture described by (2.3). What happens if (3) is relaxed?

Consider the following substitute for (3).[8]

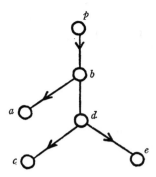

FIGURE 6

(3′) If $W^s(p)$ and $W^u(q)$ intersect at all, then there is a point of transversal intersection of $W^s(p)$ and $W^u(q)$.

With the weaker (3′) replacing (3) one still is able to prove (2.3). Moreover with either (3) or (3′), the relation $\leq$ and the diagram are invariant under perturbation. However, with a weakened version of (2.2) there is no hope of proving structural stability as the simplest counter-examples show. In fact for structurally stable $f \in \text{Diff}(M)$, (3) is satisfied.

The bulk of this section is taken from [109] with some updating, a few examples and other points added. On the other hand, many of the ideas go back quite a number of years. Certainly the local theory as mentioned in the text is of this character. Also Poincaré [89], Birkhoff [16], and M. Morse [68] all had some parts of this global picture. Since this earlier work, Andronov and Pontrjagin [6], Elsgolts [30], Peixoto [83], Reeb [94], and Thom [124] among others had made contributions toward the picture given in this section.

Besides giving simple examples of Ω-stable diffeomorphisms, the material in this section serves as an introduction to the more general theory of §I.6, where a number of these concepts have natural extensions.

I.3. **Anosov diffeomorphisms.** The examples of this section (at least roughly speaking) are at the opposite extreme from those of the preceding section in that the whole manifold consists of nonwandering points and the periodic points are dense. This is in contrast to Ω being finite as in §I.2. We give first the simplest examples of Anosov diffeomorphisms, the toral diffeomorphisms.

Consider f_0, a 2×2 matrix with integer entries and determinant ± 1, i.e., $f_0 \in \text{GL}(2, Z)$. Then f_0 can be thought of as a linear transformation of the plane R^2 which preserves the lattice L of points with integer coordinates. There is an induced diffeomorphism f of the quotient $R^2/L = T^2$, 2-dimensional torus, onto itself. This diffeomorphism $f: T^2 \to T^2$ has a fixed point p corresponding to the origin of R^2.

Now suppose f_0 is hyperbolic, for example

$$\begin{pmatrix} 1 & 2 \\ 1 & 1 \end{pmatrix}.$$

Then p will be a hyperbolic fixed point of f and the stable and unstable manifolds $W^s(p)$ and $W^u(p)$ will be the image of the eigenspaces of f_0 under the projection $\Pi: R^2 \to T^2$ (since f_0 is hyperbolic, the eigenvalues λ, μ are real and satisfy $|\lambda| > 1 > |\mu| > 0$ with 1-dimensional eigenspaces). Since $W^s(p)$ is a 1-1 immersion, it winds densely around the torus and similarly with $W^u(p)$.

The intersection points, in $W^s(p) \cap W^u(p)$ (called homoclinic points, see §I.5), are clearly dense in T^2, and it can also be shown that the periodic points of f are dense in T^2. This follows from an algebraic argument or one can use the generalized Birkhoff theorem (see I.(5.6)).

For any periodic point $q \in T^2$ of period m, the derivative of f^m at q can be thought of as $f_0^m: R^2 \to R^2$ after identifying $T_q(T^2)$ and R^2 by translation. The stable manifold $W^s(q)$ will then just be the translate of $W^s(p)$. From the Lefschetz Trace Formula (see §I.4), the number of periodic points N_m of period m is $1 - (\lambda^m + \mu^m) + \text{degree } f$. Then any g in the same homotopy class as f must have an infinite number of periodic points and therefore cannot satisfy (2.2). It turns out that f is structurally stable so that any perturbation of f' will also have periodic points dense in T^2. Everything said about f_0 extends to hyperbolic $f_0 \in GL(n, Z)$, defining what we will call toral diffeomorphisms.

The definitive version of the structural stability of f is contained in the work of Anosov [7], [8] which we will describe now.

We recall that a *Riemannian* vector space bundle E over a space X is a vector space bundle such that each fiber E_x is equipped with an inner product $(,)_x$ in a continuous manner. This allows one to speak of the norm $\|v\|$ of a vector $v \in E_x$. A *bundle map* between vector space bundles is a fiber preserving $\phi: E \to E$ of a Riemannian vector space bundle into itself and will be called *contracting* if there exists $C > 0$, $0 < \lambda < 1$ such that for all $v \in E$, $m \in Z^+$

$$\|\phi^m(v)\| < C\lambda^m \|v\|.$$

It will be called *expanding* if there exists $d > 0$, $\mu > 1$ such that for all $v \in E$, $m \in Z^+$

$$\|\phi^m(v)\| > d\mu^m \|v\|.$$

Above, we really are just using the norm in each fiber, not the inner product.

(3.1) PROPOSITION. *If X is compact, then the property of being contracting or expanding for $\phi\colon E \to E$, E a Riemannian vector space bundle over X is independent of the Riemannian metric.*

PROOF. Two norms $\| \ \|, \| \ \|'$ on fibers of E are related locally, and hence globally by $a\| \ \| \le \| \ \|' \le b\| \ \|$ for some a, $b > 0$. Thus if $\|\phi^m(v)\| \le c\lambda^m\|v\|$, then $\|\phi^m(v)\|' \le c(b/a)\lambda^m\|v\|'$.

(3.2) PROPOSITION. *The inverse of a contracting bundle automorphism is an expanding bundle automorphism and vice-versa.*

PROOF. Suppose $\|\phi^m(v)\| \le c\lambda^m\|v\|$. Then writing $\phi^m(v) = w$, $\|\phi^{-m}(w)\| \ge (1/c)(1/\lambda)^m\|w\|$.

Actually, J. Mather has shown me how to renorm the bundle so as to make $c = 1$ in the defining condition for contracting bundle automorphisms.

Since on every vector space bundle there exist Riemannian metrics, by (3.1) we can dispense with the Riemannian structure when speaking of contracting bundle maps of bundles over compact spaces.

If $f\colon M \to M$ is a diffeomorphism, then the derivative $Df\colon T(M) \to T(M)$ is a bundle automorphism of the tangent bundle of M.

Suppose now that M is Riemannian so that $T(M)$ is a Riemannian vector space bundle over M. We will say that $f\colon M \to M$ is an Anosov diffeomorphism and that M has a *hyperbolic structure* for f if the following condition is satisfied: there is a splitting of the tangent bundle $T(M)$ into a continuous (not necessarily C^r) Whitney sum $T(M) = E^s + E^u$, invariant under $Df\colon T(M) \to T(M)$ so that $Df\colon E^s + E^s$ is contracting and $Df\colon E^u \to E^u$ is expanding.

The Riemannian structure of $T(M)$ restricts to give a Riemannian structure on E^s and E^u so that this condition makes sense; furthermore, in case M is compact, by our previous comment, the Riemannian structure is unnecessary.

(3.3) THEOREM (ANOSOV) ([7], [8]). *An Anosov diffeomorphism f of a compact manifold M is structurally stable. Furthermore if there is a Lebesgue invariant measure for f on M, then the periodic points are dense and f is ergodic. Finally the Anosov diffeomorphisms are an open set in* Diff(M).

For the proof of the first statement of (3.3) see the exposition of J. Moser's proof by J. Mather in the appendix. For the last sentence, see §I.8.

It is apparent that the toral diffeomorphisms are Anosov diffeomorphisms; the splitting by f at p translates to each point of T^n to give the desired global splitting.

From an invariant measure for a diffeomorphism f of a compact manifold M, one can see easily that every point is nonwandering, i.e., $\Omega = M$. It is from this fact that Anosov concludes the density of the periodic points of f.

(3.4) PROBLEM. Is it true that for every Anosov diffeomorphism of a compact manifold M, $\Omega = M$, or equivalently, the periodic points are dense in M?[9] A second question is: does every Anosov diffeomorphism have a fixed point?

Motivation for this work of Anosov comes not only from the toral diffeomorphisms, but more importantly from geodesic flows on manifolds with negative curvature, where Anosov's ergodicity solves an old problem. This is the 1-parameter analogue of (3.3) and will be discussed later in our survey.

For (3.3), the basic idea of Anosov's proof is to construct through each point p of M, a generalized stable manifold $W^s(p)$. This will be a 1-1 immersed cell with the property that for each $x \in W^s(p)$, the tangent space $T_x(W^s(p))$ coincides with $E_x^s \subset T_x(M)$. Furthermore $f(W^s(p)) = W^s(f(p))$, and x, y are in the same $W^s(p)$ if and only if $d(f^m x, f^m y) \to 0$ as $m \to \infty$.

Although each $W^s(p)$ is smooth, $W^s(p)$ only depends continuously on p (recall that the splitting of $T(M)$ was only required to be continuous). One may think of the $W^s(p)$ giving a continuous foliation of M. The existence and basic properties of $W^s(p)$ are based on old work by Perron [88].

Theorem (3.3) states that the Anosov diffeomorphisms are an open set in $\text{Diff}(M)$. On the other hand Anosov has examples to show that this would be false if one imposed a *smooth* splitting of $T(M)$ rather than a continuous one in the definitions.

The following is a basic and beautiful unsolved problem.

(3.5) PROBLEM. Find all examples of Anosov diffeomorphisms of compact manifolds (up to topological conjugacy of course) such that $M = \Omega$.[10] What compact M admit Anosov diffeomorphisms? Must M be covered by Euclidean space?

There do exist nontoral Anosov diffeomorphisms. We will show this now and in fact give the most general known way of constructing Anosov diffeomorphisms.

Suppose that G is a connected simply connected Lie group with Lie algebra $\mathfrak{G}$ and a uniform discrete subgroup Γ (uniform means that the coset space G/Γ is compact). Suppose also that $f_0: G \to G$ is a continuous automorphism such that $f_0(\Gamma) = \Gamma$ and the derivative at the identity $f_0' : T_e(G) \to T_e(G)$ is hyperbolic (throughout this discussion it will be helpful to keep the toral case, with $G = R^n$, in mind). If $T_e(G)$ is identified with $\mathfrak{G}$ then f_0' becomes the Lie algebra automorphism induced from f_0. From this data we will construct an Anosov diffeomorphism $f : G/\Gamma \to G/\Gamma$. At this writing, this is the most general known construction of an Anosov diffeomorphism.

Since the linear automorphism $f_0' : \mathfrak{G} \to \mathfrak{G}$ is hyperbolic, we get the usual invariant splitting $\mathfrak{G} = \mathfrak{G}^s + \mathfrak{G}^u$. Furthermore, (see [114]) there exist constants c, c' such that $0 < c < 1 < c'$ and an inner product on $\mathfrak{G}$ so that

$$\|f_0'(v)\| < c\|v\| \quad \text{all } v \in \mathfrak{G}^s,$$
$$\|f_0'(u)\| > c'\|u\| \quad \text{all } u \in \mathfrak{G}^u.$$

Next by right translations, identifying $\mathfrak{G}$ with $T_e(G)$, the splitting and inner product are imposed on the tangent space of every point of G. For this Riemannian metric on G, it is easily checked that $f_0 : G \to G$ is given a hyperbolic structure or that $f_0 : G \to G$ is an Anosov diffeomorphism.

Furthermore, this splitting of $T(G)$ and the Riemannian metric are both invariant under the action of G on G given by right translation. In particular they are right invariant under Γ and so f_0 induces an Anosov diffeomorphism f on the compact coset space G/Γ.

For the existence of the f_0 in the previous construction, the next proposition shows that G must be nilpotent.

(3.6) PROPOSITION. *Suppose that $\phi: \mathfrak{G} \to \mathfrak{G}$ is a Lie algebra automorphism which is hyperbolic as a linear map. Then $\mathfrak{G}$ must be nilpotent.*

For a proof, A. Borel has given me the following reference: let $\mathfrak{G}$ be a finite dimensional Lie algebra over a field having an automorphism no eigenvalue of which is a root of unity; then $\mathfrak{G}$ is nilpotent. Exercise in Bourbaki with hints: *Algebras de Lie*, Ex. 21b, p. 124.

Now that we know that this construction forces G to be nilpotent, and that Γ is a uniform discrete subgroup, the results of Malcev [61], summarized in [12a], become important.

(3.7) THEOREM (MALCEV). (a) *A necessary and sufficient condition for a discrete group Γ to occur as a uniform subgroup of a simply connected nilpotent Lie group is that Γ be a finitely generated nilpotent group containing no elements of finite order.*

(b) *A necessary and sufficient condition on a nilpotent simply connected Lie group G that there exist a uniform discrete subgroup Γ is that* *the Lie algebra of G has rational constants of structure in some basis.*

(c) *If Γ_i is a uniform discrete subgroup of a simply connected nilpotent group G_i, $i = 1$, 2, then any isomorphism $\Gamma_1 \to \Gamma_2$ can be uniquely extended to an isomorphism $G_1 \to G_2$.*

The coset space G/Γ, G, Γ as above is called a *nilmanifold*.

While (3.6) and (3.7) give some general perspective on our class of homogeneous space Anosov diffeomorphisms, this situation cannot be said to be completely understood. There certainly do exist, however, many nontoral examples of Anosov diffeomorphisms on nilmanifolds as special cases of the above construction. We give two of them now with dim $G = 6$.

Let G_1, G_2 be copies of the three dimensional simply connected, nonabelian nilpotent Lie group. We take a basis X_i, Y_i, Z_i of $\mathfrak{G}_i$, $i = 1$, 2 with the bracket relations $[X_i, Y_i] = Z_i$, $i = 1$, 2 and all other brackets zero. The main group G of our basic construction above will be $G_1 \times G_2$. For each real number $\lambda > 1$ we define a hyperbolic automorphism f_0 of G by specifying f_0' (f_0', $\mathfrak{G}$, $\mathfrak{G}^s$, etc. as in the above construction) on $\mathfrak{G}$ in terms of the basis as follows.

EXAMPLE 1	EXAMPLE 2
$X_1 \to \lambda X_1$	$X_1 \to \lambda X_1$
$Y_1 \to \lambda^2 Y_1$	$Y_1 \to \lambda^{-3} Y_1$
$Z_1 \to \lambda^3 Z_1$	$Z_1 \to \lambda^{-2} Z_1$
$X_2 \to \lambda^{-1} X_2$	$X_2 \to \lambda^{-1} X_2$
$Y_2 \to \lambda^{-2} Y_2$	$Y_2 \to \lambda^3 Y_2$
$Z_2 \to \lambda^{-3} Z_2$	$Z_2 \to \lambda^2 Z_2$

Note that in both examples brackets are preserved. In Example 1, one sees that $\mathfrak{G}^u$, $\mathfrak{G}^s$ are both ideals which coincide with nilsubalgebras $\mathfrak{G}_1$ and $\mathfrak{G}_2$ respectively. In this case G is the product of the corresponding subgroups, $G = G^u \times G^s$.

In Example 2, both $\mathfrak{G}^u$ and $\mathfrak{G}^s$ are seen to be abelian, but they are not ideals and G is not (in the group sense) a product of the corresponding subgroups G^u and G^s.

The next step is to find a uniform discrete subgroup $\Gamma \subset G$ such that $f_0(\Gamma) = \Gamma$. For this we will use matrices with coefficients in an algebraic number field. Let $K = Q(3^{1/2})$, the number field of $3^{1/2}$ adjoined to the rationals, and $\sigma: K \to K$ the nontrivial Galois automorphism (sending $3^{1/2}$ into $-(3^{1/2})$).

We may suppose that $\mathfrak{G}_1$ and $\mathfrak{G}_2$ are each represented by matrices of the form,

$$(3.8) \qquad \begin{pmatrix} 0 & X & Z \\ 0 & 0 & Y \\ 0 & 0 & 0 \end{pmatrix} \quad X, Y, Z \in R$$

and $\mathfrak{G} = \mathfrak{G}_1 \times \mathfrak{G}_2$ becomes the space of matrices

$$\begin{pmatrix} A & 0 \\ 0 & B \end{pmatrix},$$

A, B each of the form (3.8). We will take Γ_0 then to be the lattice of $\mathfrak{G}$ of matrices of the form

$$\begin{pmatrix} A & 0 \\ 0 & A^\sigma \end{pmatrix}$$

where A is as in (3.8) but X, Y, Z are restricted to be algebraic integers in K and A^σ is the image of A under the map induced by $\sigma \colon K \to K$.

Then if $\lambda = 2 + 3^{1/2}$, $\lambda \lambda^\sigma = 1$ and f_0' preserves Γ_0. We take Γ to be the image of Γ_0 under the exponential map $\mathfrak{G} \to G$. Then it can be proved that Γ is a uniform discrete subgroup of G with $f_0(\Gamma) = \Gamma$. This finishes the description.

One can generalize the previous construction by using a diagonal process defined by the Galois automorphisms of an algebraic number field. See Weil [125] for this type of argument.

It seems possible that if $f \in \mathrm{Diff}(M)$ is Anosov, where M is compact, then M is covered by Euclidean space $\tilde{M}$, and that even the induced Anosov diffeomorphism on $\tilde{M}$ is topologically conjugate to a linear hyperbolic map. However, we have an example of an Anosov diffeomorphism $f \colon V \to V$ where V is a simply connected complete Riemannian manifold, noncompact and not Euclidean space. This example goes as follows.[11]

The construction starts with G as $SL(2, C)$ and proceeds something like our earlier nilpotent examples (but no Γ this time).

Let $\phi \colon G \to G$ be the inner automorphism obtained by conjugation with the matrix

$$A = \begin{pmatrix} \alpha & 0 \\ 0 & \alpha^{-1} \end{pmatrix}$$

in G where $\alpha > 1$. Then if $e^{\mathrm{ad}A} \colon \mathfrak{G} \to \mathfrak{G}$ is the Lie algebra automorphism, we have the invariant decomposition $\mathfrak{G} = \mathfrak{G}^s + \mathfrak{G}^u + h$ where $\mathfrak{G}^s$ is contracting under $e^{\mathrm{ad}A}$, $\mathfrak{G}^u$ expanding and h is invariant pointwise. In fact h is the Lie algebra of the centralizer H of A, of all diagonal matrices. Just as in the previous construction we put a metric on

$\mathfrak{G} = T_e(A)$ which is right translated around, but contains a degenerate component corresponding to h.

On G/H, however, the degeneracy is divided out so that we have an induced Anosov diffeomorphism ϕ_0: $G/H \rightarrow G/H$. G/H is a 4-dimensional manifold which is not contractible, but clearly simply connected.

Novikov informed me that he could prove that if $f \in \text{Diff}(M)$ is Anosov, M compact, where the dimension of $W^u(x)$ is one less than the dimension of M, then $\pi_1(M)$ is abelian and M is covered by Euclidean space.

The two dimensional toral example was first communicated to me by Thom to show that there was an open set in $\text{Diff}(T^2)$ of diffeomorphisms with no contracting periodic points, therefore implying that diffeomorphisms satisfying (2.2) were not dense. After adding some geometry to the example, I showed it to Anosov when I spoke on the examples of §I.5 in the Soviet Union in 1961. By 1962 Anosov announced his theorem on structural stability in the context of what is called here Anosov diffeomorphisms. Proofs have now appeared [9].

The problem of the existence of (compact) nontoral diffeomorphisms was posed by Anosov in his Congress talk, Moscow 1966. Previously, after putting this problem into Lie group perspective, I had consulted many Lie group experts to arrive finally at what is here. In particular, conversations with Boothby, Borel, Hochschild, and Langlands were very helpful. The 6-dimensional Example 2 as well as the explicit algebraic number theory approach were given to me by Borel.[12]

I.4. The zeta function of a diffeomorphism.

Suppose f: $M \rightarrow M$ is a diffeomorphism with the property that $N_m < \infty$, $m = 1, 2, \cdots$ where $N_m = N_m(f)$ is the number of fixed points of f^m. This is a generic property (see §I.6). Then following Artin-Mazur [12], one defines the zeta function of f as the formal power series $\zeta(t) = \exp \sum_{m=1}^{\infty} (1/m) N_m t^m$. This turns out to be an interesting invariant of f. Of course $\zeta_f(t) = \zeta(t)$ is an invariant of the topological conjugacy class of f and even of the conjugacy class "on Ω" of f.

The zeta function thus contains all the information about the numbers $N_m = N_m(f)$ where N_m counts all the periodic points of period m. But this is different from $K_m = K_m(f)$ which denotes the number of periodic points of least period m. The number K_m is more directly interesting in many respects and it is natural to ask for the relation between N_m and K_m. From the definition it follows directly that

(4.1) PROPOSITION.

$$\sum_{l \ divides \ m} K_l = N_m.$$

Narasimhan pointed out to me that one solves (4.1) for the K_l by the Mobius inversion theorem (see [36]). This gives

(4.2) PROPOSITION.

$$K_m = \sum_{l \ divides \ m} \mu(l) N_{m/l}.$$

Here if $l = p_1 \cdots p_r$ where the p_i are distinct primes, then $\mu(l) = (-1)^r$, $\mu(1) = 1$, and if l contains a power of a prime, $\mu(l) = 0$. The function $\mu(l)$ is called the Möbius function.

Observe that m always divides K_m (i.e., $K_m/m \in Z^+$).

The inspiration for the above zeta function is the Weil zeta function of an algebraic variety over a finite field. Dwork recently proved the rationality of this zeta function, see [101] for a general reference.

For the differentiable version, there is the following theorem.

(4.3) THEOREM (ARTIN-MAZUR [12]). *For any compact manifold, there is a dense set of* Diff(M) *for which the following estimate holds:*

$$N_m \leqq Ck^m.$$

Here C, k are positive constants which depend only on the diffeomorphism f and $N_m = N_m(f)$.

(4.4) COROLLARY. *For a dense set of* Diff(M), *the zeta function has a positive radius of convergence, so it can really be considered a function.*

Actually Artin and Mazur define N_m to be the number of isolated fixed points of f^m, while permitting f^m to have an infinite number of fixed points. Thus, for example, they do not know whether the fixed point set is finite for the maps in the dense set they obtain.

The proof of (4.3) uses algebraic approximation techniques which go back to John Nash [73]. Actually Artin and Mazur define $\zeta(t)$ for differentiable maps for which $N_m < \infty$ and prove their theorem in the more general context of differentiable maps. The following problem then becomes important.

(4.5) PROBLEM. Is $\zeta(t)$ generically rational (i.e., is ζ_f rational for a Baire set of f)?[13]

This goes beyond their theorem in two ways. First, generically true means true for a Baire set which, of course, is much bigger than simply a dense set. Secondly, rationality is stronger than possessing a positive radius of convergence. Rationality is especially important because this means for the diffeomorphism that the poles and zeros of the zeta function, a finite number of invariants, determine the infinite set of N_m. The N_m are of course very important objects to get ones hands on.

In the direction of (4.5), Artin and Mazur [12] asked if diffeomorphisms in their dense set have a zeta function which is algebraic.

More recently, there has been proved the following

(4.6) THEOREM (K. MEYER). *If $f \in \mathrm{Diff}(M)$, M compact, satisfies Axiom A (see §1.6), then the estimate of (4.3) is valid.*

K. Meyer's proof of this is very simple and if one had the density (see §1.6) for Axioms A and B, this would of course supersede (4.3). We will now examine the zeta function for our examples.

If $f \colon M \to M$ is a diffeomorphism such that $N_m < \infty$ for all $m \in Z$ and Λ is a closed invariant subset of M, then by definition $\zeta_\Lambda(f)$ $= \exp \sum_{m=1}^{\infty} (1/m) N_m' t^m$ where N_m' is the number of $x \in \Lambda$ such that $f^m(x) = x$.

(4.7) PROPOSITION. *Suppose for $f \in \mathrm{Diff}(M)$, the periodic points are all contained in the union of two disjoint closed invariant subsets Λ_1, Λ_2 of M. Suppose also that ζ_{Λ_1} and ζ_{Λ_2} are rational (or convergent). Then $\zeta_f = \zeta$ is rational (or convergent) and in fact $\zeta(t) = \zeta_{\Lambda_1}(t) \cdot \zeta_{\Lambda_2}(t)$.*

PROOF.

$$\zeta(t) = \exp \sum \frac{N_m' + N_m''}{m} t^m = \exp \sum \frac{N_m'}{m} t^m \exp \sum \frac{N_m''}{m} t^m = \zeta_{\Lambda_1}(t) \cdot \zeta_{\Lambda_2}(t).$$

(4.8) LEMMA FROM CALCULUS. $\log (1/(1-y)) = \sum_{k=1}^{\infty} (1/k) y^k$.

For the diffeomorphisms of (2.2) the following theorem gives the zeta function.

(4.9) THEOREM. *Suppose for $f \in \mathrm{Diff}(M)$, Ω is finite. Then clearly $\Omega = \bigcup_{\gamma \in P} \Omega_\gamma$ where P is the set of periodic orbits of f, and Ω_γ is the set of points of Ω in γ. The zeta function of f is the following, where $m(\gamma)$ = period of γ,*

$$\zeta(t) = \prod_{\gamma \in P} \frac{1}{1 - t^{m(\gamma)}}.$$

PROOF. By (4.8) $1/(1 - t^m(\gamma)) = \exp \sum_{k=1}^{\infty} (1/k) t^{m(\gamma)k}$. Apply (4.7). The Lefschetz number $L(p) = L(p, f)$ of a hyperbolic fixed point p

of $f \in \mathrm{Diff}(M)$ may be defined most simply, perhaps, as ± 1 where the sign is the sign of $\det(I - Df(p))$, $I: T_p(M) \to T_p(M)$ being the identity.

A modern proof of a more general version of the following theorem of Lefschetz may be found in Dold [26].

(4.10) LEFSCHETZ TRACE FORMULA. Suppose $f \in \mathrm{Diff}(M)$ has only hyperbolic fixed points and $\mathrm{Fix}(f)$ denotes the set of fixed points of f. Then

$$\sum_{p \in \mathrm{Fix}(f)} L(p) = \Lambda(f) \quad \text{where}$$

$$\Lambda(f) = \sum_{i=0}^{\dim M} (-1)^i \, \mathrm{Trace}(f_{*i}: H_i(M, R) \to H_i(M, R)).$$

Here f_{*i} is the induced automorphism of the ith homology group of M with real coefficients.

The following proposition follows from the definition of $L(p)$ and the eigenspace decomposition of $Df(p)$. (One may assume $Df(p)$ to be semisimple in the proof.)

(4.11) PROPOSITION. *For* $p \in \mathrm{Fix}(f)$, $f \in \mathrm{Diff}(M)$, $L(p) = (-1)^u \Delta$, *where* $u = \dim W^u(p)$ *and* $\Delta = +1$ *if* f *preserves orientation on* $W^u(p)$ *and* $\Delta = -1$ *if* f *reverses it.*

The following is well known and will be useful in computing the zeta function for some of the Anosov diffeomorphisms.

(4.12) PROPOSITION. *Suppose* $f \in \mathrm{Diff}(M)$ *is such that for every* $m \in Z^+$ *and every* $x \in \mathrm{Fix}(f^m)$, $L(x, f^m) = +1$. *Then* $\zeta(t) = \prod_{i=1}^{\dim M} \zeta_i(t)^{(-1)^i}$ *where*

$$\zeta_i(t) = \prod_j (1 - \lambda_{ij}t)^{-1} \quad and \quad \lambda_{ij}, j = 1, \cdots, \dim H_i(M, R),$$

are the eigenvalues (generalized and counted with multiplicity) of $f_{*i}: H_i(M, R) \to H_i(M, R)$.

PROOF. By (4.10)

$$N_m(f) = \sum_{i=0}^{\dim M} (-1)^i \, \mathrm{Trace}(f^m)_{*i}, \qquad m = 1, 2, 3, \cdots$$

$$= \sum_{i=0}^{\dim M} (-1)^i \sum_{r=1}^{\dim H_i} \lambda_{ri}^m.$$

So we obtain

$$\log \zeta(t) = \sum_{m=1}^{\infty} \frac{1}{m} \sum_{i=0}^{\dim M} (-1)^i \sum_{r=1}^{\dim H_i} \lambda_{ri}^m t^m$$

or

$$\zeta(t) = \prod_{i=0}^{\dim M} \zeta_i(t)^{(-1)^i}$$

where

$$\log \zeta_i(t) = \sum_{m=1}^{\infty} \frac{1}{m} \sum_{r=1}^{\dim H_i} (\lambda_{ir}t)^m$$

$$= \sum_{r=1}^{\dim H_i} \sum_{m=1}^{\infty} \frac{1}{m} (\lambda_{ir}t)^m$$

$$\log \zeta_i(t) = \sum_{r=1}^{\dim H_i} \log(1 - t\lambda_{ir})^{-1} \text{ by } (4.8)$$

and

$$\zeta_i(t) = \prod_{r=1}^{\dim H_i} (1 - t\lambda_{ir})^{-1}.$$

For any $f \in \text{Diff}(M)$, the function $\zeta(t)$ defined in (4.12) is well defined even though $L(x, f^m)$ is not always 1. It will be called the false zeta function of f and denoted by $\tilde{\zeta}(t)$ or $\tilde{\zeta}_f(t)$. It is rational and its expansion counts the periodic points algebraically. In fact, the whole difficulty of the problem of the rationality of the (honest) zeta function is that it counts the periodic geometrically, not algebraically. Proposition (4.12) shows that under the condition $L(x, f^m) = 1$ for all $x \in \text{Fix}(f^m)$, and $m \in Z^+$, the false and honest zeta functions coincide.

Note that if $\tilde{N}_m$ is the number of points of f of period m counted algebraically, i.e., $\tilde{N}_m = \sum_{x \in \text{Fix}(f^m)} L(p, f^m)$, then (4.12) shows that $\tilde{\zeta}(t) = \exp \sum_{m=1}^{\infty}(1/m)\tilde{N}_m t^m$ and one can see how the following theorem of Fuller [31] fits into this context (see also [38]).

(4.13) THEOREM. *Suppose $h: L \to L$ is a homeomorphism of a polyhedron of nonzero Euler characteristic. Then h has a periodic point.*

Otherwise all the $\tilde{N}_m$ would be zero and $\tilde{\zeta}$ would be one. But the degree of $\tilde{\zeta}$ is minus the Euler characteristic (from (4.12)).

(4.14) PROPOSITION. *Suppose $f: M \to M$ is an Anosov diffeomorphism such that the corresponding expanding bundle E^u is orientable. Then ζ_f is rational and*

(a) *if Df: $E^u \to E^u$ is orientation preserving then*

$$\zeta_f = \tilde{\zeta}_f \quad \text{if dim fiber } E^u \text{ is even,}$$

$$\zeta_f = 1/\tilde{\zeta}_f \quad \text{if dim fiber } E^u \text{ is odd,}$$

(b) *if Df: $E^u \to E^u$ is orientation preserving then*

$$\zeta_f(t) = \tilde{\zeta}(-t) \quad \text{if dim fiber } E^u \text{ is even,}$$

$$\zeta_f(t) = 1/\tilde{\zeta}(-t) \quad \text{if dim fiber } E^u \text{ is odd.}$$

This follows directly from (4.11) and (4.12).

It seems likely that looking at a double covering of M, one could furthermore prove that the zeta function of every Anosov diffeomorphism was rational.

For the toral case of §I.3 defined by hyperbolic $f_0 \in \mathrm{GL}(n, Z)$, one finds the zeta function defined explicitly in terms of the eigenvalues of f_0. In this case f_0 coincides with the automorphism of $H^1(T^n, Z)$ induced by $f\colon T^n \to T^n$. By the Kunneth formula the whole cohomology ring of T^n is given as a tensor product of $H^*(T^1)$ and so one obtains all of the eigenvalues of $f^*\colon H^*(T^n) \to H^*(T^n)$ as products of the eigenvalues of f_0. One thus obtains easily via (4.14)

(4.15) PROPOSITION. *For the toral diffeomorphism $f\colon T^n \to T^n$ defined by hyperbolic $f_0 \in \mathrm{GL}(n, Z)$ with the eigenvalues $\lambda_1 \cdots \lambda_n$ of f_0, we have:*

(a) $\Lambda(f^m) = \prod_i (1 - \lambda_i^m)$,

(b) $\tilde{\zeta}(t) = \prod (1 - \lambda_{i_1} \lambda_{i_2} \cdots \lambda_{i_k} t)^{(-1)^{k+1}}$,

$$\text{all } (i_1, \cdots, i_k) \ni 1 \leqq i_1 < i_2 < \cdots < i_k \leqq n,$$

(c) *$\zeta(t)$ is defined from $\tilde{\zeta}(t)$ according to (4.14) where one checks the appropriate case from the λ_i with $|\lambda_i| > 1$.*

Finally, we remark that through the work of Matsushima [63], Mal'cev [61], Nomizu [75], and Kostant [54], one can compute the zeta functions for the nilpotent examples of §I.3 quite explicitly.

I.5. **Shift automorphisms and homoclinic points.** From the preceding sections, one might ask whether the set Ω of nonwandering points must be a manifold generically (allowing certainly for components to have varying dimensions). The examples of §§I.2 and 3 have this property. Here we will see that the answer is no, and in fact give an example of a diffeomorphism of S^2, Ω-stable, such that Ω is the union of a Cantor set and two isolated points.

First a description of the shift automorphism of symbolic dynamics will be given (see [14] or [35a] for more details). Let S be a finite set, discrete topology, consisting of N elements and define X_S to be the set of functions from Z to S provided with the compact open topology (Z has the discrete topology). If $a \in X_S$, the value of a at $m \in Z$ will be denoted by a_m and we write $a = \prod a_m$. Then a may be thought of as a doubly infinite sequence of elements of S with a decimal point between a_0 and a_1, thus a $a = (\cdots a_{-1}a_0 \cdot a_1 a_2 \cdots)$. An important special case is where S has two elements and here we could assume each a_i is either 0 or 1. For general S, X_S is homeomorphic to a Cantor set.

Define a map $\alpha: X_S \to X_S$ by $(\alpha(a))_m = a_{m+1}$. In terms of the doubly infinite sequences, α shifts the decimal point one place to the right. It is easily seen that α is a homeomorphism, called the shift automorphism of X_S. It has been widely studied in ergodic theory and probability [14] as well as topological dynamics [35a].

(5.1) PROPOSITION. *The periodic points of α are dense in X_S and if C_k is the number of periodic points of period k (i.e. fixed points of α^k) $k > 0$ then $C_k = N^k$ where N is the cardinality of S.*

PROOF. The element $a = \prod a_m \in X_S$ will be periodic of period k precisely when $a_m = a_{m+k}$ for all $m \in Z$. Thus it is determined by $a_1, \cdots, a_k$ with $a_1, \cdots, a_k$ arbitrary elements of S. Given any $b = \prod b_m \in X_S$ and K large, one can choose a periodic approximation $a = \prod a_m$ of b with $a_i = b_i$ for $|i| < K$. The proposition follows.

(5.2) COROLLARY. *The zeta function for $\alpha: X_S \to X_S$ can be defined as in §I.4 and in fact $\zeta(t) = 1/(1 - Nt)$ where $N = $ cardinality of S.*

This follows from (5.1) with the aid of (4.8).

M. Morse has proved (see [35a]) that there is a subset of X_S, homeomorphic to a Cantor set, which is a minimal set for α.

To see how symbolic dynamics enters into our diffeomorphism problem, we will first describe an example of a diffeomorphism g mapping a subset Q of the plane into the plane. Here $g(Q)$ is not a subset of Q, but eventually we will use g to define a global diffeomorphism f of S^2 onto itself. One might think of Q as a neighborhood (not invariant) of an indecomposable piece of the nonwandering points of this $f: S^2 \to S^2$.

Take then Q to be a square in the plane R^2, for example, $Q = \{ (x, y) \in R^2 \mid |x| \leq 1, |y| \leq 1 \}$. Then g will map Q into the region bounded by dotted lines with $g(A) = A'$ etc. in Figure 1.

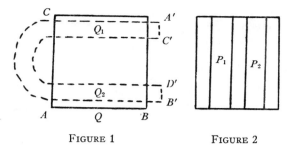

FIGURE 1   FIGURE 2

We will take any such g which has the following properties.

(a) g is a diffeomorphism of Q onto the region in Figure 1 bounded by dotted lines sending $A \to A'$, $B \to B'$ etc.

(b) on each component P_1, P_2 of $g^{-1}(g(Q) \cap Q)$, g will be a linear map (up to a translation).

To understand (b) note that as a consequence of it, P_1, P_2 will be as in Figure 2 and $g(P_i) = Q_i$, $i = 1, 2$.

The reader will be able to verify that the intersections of all the images $g^m(Q)$, $m = 1, 2, \cdots$ or, more accurately, $\bigcap_{m=1}^{\infty} g^m(Q^{(m)})$ where $Q^{(m)} = Q \cap$image g^{m-1}, is a product of a Cantor set and the interval $|x| \le 1$.

Define Λ to be the intersection, $\bigcap_{m \in Z} g^m(Q^{(m)})$, $Q_0 = Q$, $Q^{(m)}$ as above for $m > 0$ and for $m < 0$, $Q^{(m)} = g^m(Q^{(m+1)})$. Thus Λ may be thought of as the set of nonwandering points of $g: Q \to R^2$.

The careful reader will be able to check for himself the next proposition (which is in [115]).

(5.3) PROPOSITION. *The subset Λ of Q is compact, invariant under g, indecomposable and on Ω, g is topologically conjugate to the shift automorphism $\alpha: X_S \to X_S$, with the cardinality of $S = 2$.*

Furthermore one can prove stability with the less obvious proposition [115].

(5.4) PROPOSITION. *For a perturbation g' of g, Λ' defined similarly is also compact and invariant under g'. Then $g': \Lambda' \to \Lambda'$ is also topologically conjugate to the shift $\alpha: X_S \to X_S$.*

Thus (at least after we globally extend g) we have another example of a stable indecomposable piece of nonwandering points.

One may modify the above example in the following way. The image $g(Q)$ may wind half-way around Q before passing through Q the second time, or even wind around Q several times for that matter (Figures 3 and 4). This won't change $g: \Lambda \to \Lambda$, but g will be different on $U(\Lambda)$ where U is any neighborhood of Λ. The intrinsic picture (with respect to Ω) is the same for Figures 1, 3, 4 but they differ extrinsically (in any neighborhood of Ω).

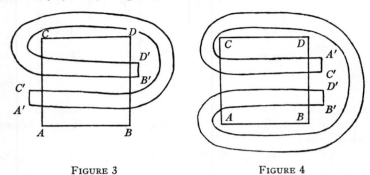

FIGURE 3   FIGURE 4

One may further modify the above examples by having $g(Q)$ passing through Q several times (see Figures 5 and 6).

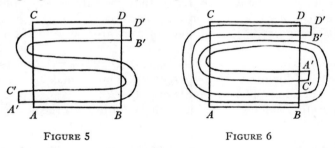

FIGURE 5   FIGURE 6

In all of these examples it is important to keep the linearity condition (b) above. Then one may define and analyze Λ as in the first case. Always $g: \Lambda \to \Lambda$ will be topologically conjugate to a shift automorphism $\alpha: X_S \to X_S$ and stably so. The cardinality of S will equal the number of components in $Q \cap g(Q)$, e.g., three for Figure 5 and four for Figure 6. Thus all of the shift automorphisms occur in the above framework.

To really complete this picture, Λ above must appear as an indecomposable piece of the nonwandering points of a global diffeomorphism. We construct such an $f: S^2 \to S^2$ now which extends the map $g: Q \to R^2$ of Figure 1. Consider Figure 7.

We have put the square Q into a disk $D^2 \subset R^2$ and we extend g to $g_0: D^2 \to D^2$ by mapping G diffeomorphically onto G' and F onto F'. The map $g_0: F \to F'$ is defined so that it is a contraction about some fixed point p_0 in F'. This $g_0: D^2 \to D^2$ will be a diffeomorphism of D^2 onto a subset of D^2 so that the nonwandering set is the disjoint union of Λ and p_0. Finally one easily extends g_0 to $f: S^2 \to S^2$ so that the non-

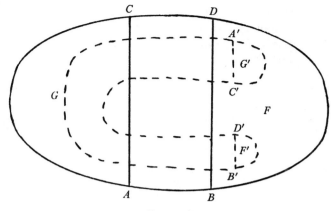

FIGURE 7

wandering points $\Omega = \Omega(f) = \Lambda \cup p_0 \cup q_0$ where q_0 is an expanding fixed point of f outside of D^2. This f is our desired global diffeomorphism.

At this point it seems appropriate to give a general way of constructing Ω-stable diffeomorphisms of S^2.

Take any diffeomorphism $f: S^2 \to S^2$ satisfying (2.2) with a contracting fixed point p. Let V be a contracting disk neighborhood of p and redefine f on V to be g_0 as described above (the "surgery" of §I.10). More generally let $f: S^2 \to S^2$ satisfy (2.2) with a contracting periodic orbit $p_1, \cdots, p_k$ and let V be a disk neighborhood of p_1 such that f^k contracts V into its interior. Then one modifies f (via surgery again) on $\bigcup_{i=0}^{k-1} f^i(V)$ to obtain $f': S^2 \to S^2$ so that on V, f'^k is conjugate to g_0 above.

Finally a straightforward modification of the previous construction allows one to introduce into any diffeomorphism $f: S^2 \to S^2$ satisfying (2.2), indecomposable pieces Λ topologically conjugate to shift automorphisms on N symbols (cardinality $S = N$) where N can be anything we like.

In all of these examples the indecomposable pieces of Ω are shift automorphisms, finite periodic orbits or products of the two. We see easily from previous remarks that the zeta functions of these $f: S^2 \to S^2$ are finite products of factors of the form $1/(1 - Nt^q)$ where N and q are positive integers.

It should be noted that the Lefschetz Trace Formula imposes conditions on what products of the above form can occur in these zeta functions. It restricts the N_i, p_i that can occur in $\zeta(s) = \prod_{i=1}^{k} (1 - N_i t^{p_i})^{-1}$.

One can see an analogy between the shift automorphism and the nilmanifold examples of §I.3 by considering the shift automorphism in the following light.

Let Z_n be the cyclic group with n elements and for each $m \in Z$ let G_m be the abelian group of formal power series (starting at m), $f(x) = \sum_{i=m}^{\infty} a_i x^i$ with $a_i \in Z_n$. Put the structure of a compact group on G_m with the product topology. Define the locally compact group of all power series by $G = \bigcup_{m \in Z} G_m$. The map $\phi^s \colon G \to G$ defined by $f \to xf$ is a contraction while $\phi^u \colon G \to G$ defined by $f \to x^{-1}f$ is an expansion.

It is easily checked that the subgroup Γ of $G \times G$ defined by $\Gamma = \{ (f, \bar{f}) \,|\, f(x)$ a polynomial in G, $f(x) = \sum_{i=m}^{k} a_i x^i$, $\bar{f}(x) = \sum a_{-i} x^i \}$ is uniform (compact quotient) and discrete. The "hyperbolic" automorphism $\phi^s \times \phi^u \colon G \times G \to G \times G$ preserves Γ and the induced homeomorphism $\phi \colon G/\Gamma \to G/\Gamma$ is precisely the shift automorphism on n symbols.

One may identity $\phi^s \colon G \to G$ above with $g \colon \tilde{W}^s(p) \to \tilde{W}^s(p)$ where $p \in \Lambda$, Λ as in the example of Figure 1, $n = 2$, $\tilde{W}^s(p) = W^s(p) \cap \Lambda$.

There is a very close relation between the shift automorphisms discussed above and what are called homoclinic points, first discovered (in the restricted 3-body problem) and named by Poincaré [90].

A *homoclinic* point of $f \in \mathrm{Diff}(M)$ is a point of intersection $x \in W^s(p) \cap W^u(q)$. If $W^s(p)$ and $W^u(q)$ are transversal at x, then x will be called a *transversal* homoclinic point.

As realized by Poincaré [90], homoclinic points complicate the orbit structure of a diffeomorphism considerably. The orbit of a homoclinic point consists (clearly) of homoclinic points. Taking the case $p = q$, one sees that the existence of homoclinic x forces $W^u(p)$ to double back on itself oscillating faster and faster as it does so. For example, for the plane, we will obtain behavior something like that described in Figure 8.

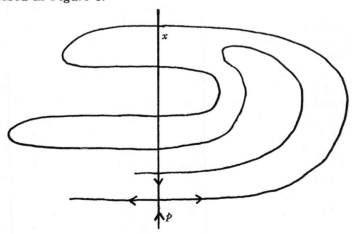

FIGURE 8

28

This is the same phenomena that is occurring in Figure 1, but looked at in a different way. In fact, the best way to understand what is going on in Figure 8 is to imbed it (in some sense) in Figure 1. A great advantage of the horseshoe approach of Figure 1 is that one gets a satisfactory picture of the orbit structure and stability while a given homoclinic point at first glance seems to defy analysis. That is the idea behind the following theorem [115].[14]

(5.5) THEOREM. *Suppose x is a transversal homoclinic point of $f \in \mathrm{Diff}(M)$. Then there is a Cantor set $\Lambda \subset M$, $x \in \Lambda$, and $m \in Z^+$ such that $f^m(\Lambda) = \Lambda$ and f^m restricted to Λ is topologically a shift automorphism.*

By (5.1) this implies:

(5.6) COROLLARY. *In every neighborhood of a transversal homoclinic point of $f \in \mathrm{Diff}(M)$, there is a periodic point.*

We interpolate a little curiosity. Note that for the shift on λ symbols, $N_m = \lambda^m$ so that by (4.2) and the fact that $K_m/m \in Z^+$, $K_m = \sum_{l/m} \mu(l)\lambda^{m/l} \equiv 0 \bmod m$ for every λ, $m \in Z^+$. This number theoretic identity for m a prime becomes $\lambda^p \equiv \lambda \bmod p$, or Fermat's Theorem.

The material in this section is mainly taken from [115] with a number of examples and figures added. The shift automorphism goes back to Hadamard (but it is even sometimes called the Bernoulli automorphism!) who used it to study geodesic flows on 2 manifolds of constant negative curvature [40]. M. Morse [69] obtained further results in the same context.

G. D. Birkhoff [17], [18] in his works on surface diffeomorphisms, studied homoclinic points. In [17], Birkhoff proved (5.6) in dimension 2, and in his [18, p. 184] he noted a resemblance between his homoclinic points and Hadamard's shift automorphism.

I came across the "horseshoe" of Figure 1 when I was trying to get a geometric picture of a variant of van der Pol's equation in N. Levinson's paper [58].[15] He had written me earlier that this equation had stably an infinite number of periodic solutions. The "horseshoe" was the first example of a structurally stable (or Ω-stable) diffeomorphism with an infinite number of periodic points [113].

Putting the shift automorphism into the group theoretic framework was done with the aid of Cal Moore.

I.6. Unification.[16] The work of the present section is motivated by the search for unity in the examples and phenomena of the preceding part of the paper. Anosov's work on hyperbolic structures on manifolds gives a clue on how to proceed. The Anosov diffeomorphisms, however, are rare among all diffeomorphisms; only certain manifolds even permit them. One is looking for a class of diffeomorphisms which include all of the previous examples in a transparent way and will at least have the possibility of including an open dense subset of $\text{Diff}(M)$ for each compact manifold M. This is provided by diffeomorphisms described now, i.e., those satisfying Axioms A and B below.

Suppose then $f: M \to M$ is a diffeomorphism of a Riemannian manifold and $\Lambda \subset M$ is a closed invariant subset. We will say that Λ is *hyperbolic* (or has a hyperbolic structure) if the tangent bundle of M restricted to Λ, $T_\Lambda(M)$ has an invariant (continuous) splitting under $Df: T_\Lambda(M) \to T_\Lambda(M)$, $T_\Lambda(M) = E^u + E^s$ such that $Df: E^s \to E^s$ is contracting and $Df: E^u \to E^u$ expanding (see §I.3 for these definitions). The dimension of the fiber of E^s need not be constant but only locally so. Since Λ is invariant, Df is an automorphism of the bundle $T_\Lambda(M)$ and this with the Riemannian metric gives sense to the above definition. Note that if Λ (or M) is compact, one may dispense with the Riemannian metric by (3.1).

The simplest examples of hyperbolic sets for diffeomorphisms are first of all the hyperbolic fixed points (§I.2) and the hyperbolic periodic points. The finite union of these cover the case of Λ finite. Next, of course, the Anosov diffeomorphisms of §I.3 are examples where the whole compact manifold is hyperbolic. Also for the examples of §I.5, the Λ homeomorphic to a Cantor set is easily checked to have a hyperbolic structure. In all of the above examples the hyperbolic sets consist of nonwandering points and the periodic points are dense in each of them (up to the unsolved problem (3.4)). The following is an example to show that hyperbolic sets need not satisfy either of these properties.

Take a diffeomorphism satisfying (2.2) which has a heteroclinic point $x \in W^s(p) \cap W^u(q)$. The 2-dimensional example of Figure 5, §I.2, will do. Thus Ω is hyperbolic and one may extend this hyperbolic structure to the orbit of x. In fact the tangent spaces of $W^s(p)$ and $W^u(q)$ at x give the desired splitting at x and similarly for each point in the orbit of x. The orbit of x together with Ω is a closed invariant set and this gives the example. The closure of the orbit of x is an indecomposable hyperbolic set.

Recall that a homeomorphism $h: X \to X$ is said to be *topologically transitive* if there is a dense orbit. Then the dense orbits form a Baire set of X (assuming that X is a compact metric space).

We will now consider a diffeomorphism $f: M \to M$ of a compact manifold which satisfies the following two properties [116].

(6.1) AXIOM A: (a) *the nonwandering set Ω is hyperbolic.* (b) *the periodic points of f are dense in Ω.*

Of these two properties (a) is the most important in what follows. In fact (b) may even be a consequence of (a).[17] The above example however shows that a proof of (a) $\Rightarrow$ (b) must use the fact that Ω consists of nonwandering points.

(6.2) THEOREM (SPECTRAL DECOMPOSITION OF DIFFEOMORPHISMS) [117]. *Suppose $f: M \to M$ satisfies (6.1). Then there is a unique way of writing Ω as the finite union of disjoint, closed, invariant indecomposable subsets (or "pieces") on each of which f is topologically transitive:*

$$\Omega = \Omega_1 \cup \cdots \cup \Omega_k.$$

(6.3) COROLLARY. *If $f: M \to M$ is as above one can write M canonically as a finite disjoint union of invariant subsets $M = \cup_{i=1}^{k} W^s(\Omega_i)$ where $W^s(\Omega_i) = \{ x \in M \mid f^m(x) \to \Omega_i, \ m \to \infty \}$.*

As the remarks at the beginning of this section indicate, the examples of §§I.2, 3 and 5 satisfy (6.1).

The spectral decomposition theorem gives a little perspective on the question of rationality of the zeta function. The zeta function of such an f will be a product of zeta functions, one for each Ω_i. It seems plausible to me that each of these zeta functions is rational.[18] The results of §§I.4 and I.5 are consistent with this.

We explain why we use the words "Spectral Decomposition for Diffeomorphisms" in (6.2). The decomposition of the manifold into invariant sets of the diffeomorphism is quite analogous to the decomposition of a finite dimensional vector space into eigenspaces of a linear map. In one case we are considering automorphisms in the category of differential topology, in the other, finite dimensional vector spaces.

As Jacques Tits pointed out, one may make this more precise by actually putting a linear transformation (generically) into the framework of (6.2). Suppose then $u: V \to V$ is a linear transformation of a complex n-dimensional vector space. By multiplying by a constant, we may suppose u has determinant 1, i.e., $u \in SL(n, C)$. We will furthermore suppose that the eigenvalues $\lambda_1, \cdots, \lambda_n$ of u have distinct absolute values which are not one. Consider the induced diffeomorphism of projective space $u_0: P^{n-1}(C) \to P^{n-1}(C)$ defined on coordinates by $Z_i \to \lambda_i Z_i$. Then u_0 will satisfy (2.2) with Ω consisting of n fixed points $(0 \cdots 0 1\ 0 \cdots 0)$. The two ways of looking at the spectral decomposition coincide.

Now we state the second of our two main axioms (introduced in [117]).

(6.4) AXIOM B.[19] *Suppose that* $f \in \mathrm{Diff}(M)$ *satisfies Axiom A and that* Ω_i, $W^s(\Omega_i)$, *etc. are as in* (6.1), (6.2), (6.3). *Then if* $W^s(\Omega_i)$ $\cap W^u(\Omega_j) \neq \varnothing$, *there exist periodic points* $p \in \Omega_i$, $q \in \Omega_j$ *such that* $W^s(p)$ *and* $W^u(q)$ *have a point of transversal intersection.*

The following generalizes the theorem of Palis, see [82] and §I.2.

(6.5) THEOREM.[20] *The set of* $f \in \mathrm{Diff}(M)$ *which satisfy Axioms* A *and* B *are open and such* f *are* Ω-*stable.*

Assuming f, Ω_i, etc. as above, we say that $\Omega_i \leq \Omega_j$ if $W^s(\Omega_i)$ $\cap W^u(\Omega_j) \neq \varnothing$. Then we also have (generalizing theorems of §2)

(6.6) THEOREM. *If* $f \in \mathrm{Diff}(M)$ *satisfies Axioms* A *and* B, *then* $\leq$ *is a partial ordering which is preserved under perturbation.*

These theorems (6.2), (6.3), (6.5), (6.6) have no proofs in the literature, but we will try to give a good sketch of their proofs in §§I.7 and 8, §I.7 for (6.2) and (6.3), §I.8 for (6.5) and (6.6).

We say that $\Omega_{i_1} \leq \Omega_{i_2} \leq \Omega_{i_3} \leq \cdots \leq \Omega_{i_n}$ is a *maximal chain* if the Ω_{i_j} are distinct, and n is maximal.

For every $f \in \mathrm{Diff}(M)$ satisfying Axioms A and B, we define the diagram $\Delta(f)$ as follows. $\Delta(f)$ is a linear graph whose vertices correspond to the Ω_i, labeled by conjugacy class, and directed 1-simplices join consecutive vertices of maximal chains. The diagram $\Delta(f)$ is invariant under perturbations of f. Generalizing problem (2.4) is

(6.6)a PROBLEM.[21] What diagrams can occur for diffeomorphisms satisfying Axioms A and B? Given first the manifold M, what diagrams can occur for $f \in \mathrm{Diff}(M)$ satisfying Axioms A and B? Finally one can label the diagrams with conjugacy classes of germs of diffeomorphisms on neighborhoods of the Ω_i as in §I.2 and ask the above two questions for these *labeled diagrams*.

One can see that a prototype of diffeomorphisms satisfying Axioms A and B are those of §I.2 with Axiom 3 replaced by Axiom 3' there. The above results may be construed as saying that we have succeeded in relaxing the hypothesis that Ω is finite. The diagram here gives sort of a very generalized gradient structure to these diffeomorphisms.

The main point of Axioms A and B and subsequent theorems is that the hypotheses include and unify all known Ω-stable diffeomorphisms, while describing an open set of $\mathrm{Diff}(M)$ which is amenable to study. In fact the above theorems as well as those in the future sections give the beginnings of a structure theory for diffeomorphisms satisfying Axioms A and B.

Thus the question, "are these diffeomorphisms dense in $\text{Diff}(M)$," becomes particularly sharp. This is not yet settled. In this direction, the first theorem is

(6.7) THEOREM. *For compact M, the following properties of $f \in \text{Diff}(M)$ are generic:*
  (a) *Every periodic point is hyperbolic.*
  (b) *For each pair of periodic points p, $q \in M$, $W^s(p)$ and $W^u(q)$ have transversal intersection.*

This is proved in [55] and [114]. In [86] there is a polished version, which also proves the noncompact case. In [2] there is an account done in the general framework of transversality theory.

Note that if f satisfies (6.7)(a) then for each $m \in Z^+$, the number of periodic points of period m is finite, i.e., $N_m < \infty$.

The other main approximation theorem is related to Pugh's C^1 solution of the problem of the "closing lemma" [91]. This can be stated as follows.

(6.8) THEOREM (PUGH). *Suppose $f \in \text{Diff}(M)$, and $x \in M$ is recurrent in the sense that $\phi: Z \to M$ defined by $\phi(m) = f^m(x)$ is not a homeomorphism onto its image. Then there is a C^1 approximation f' of f such that x becomes a periodic point of f'.*

Pugh uses the methods of (6.8) to prove the following [92].

(6.9) THEOREM. *For compact M, the property (6.1)(b) is generic in the C^1 sense. In other words suppose we put the C^1 topology on $\text{Diff}(M)$ and let G be the set of $f \in \text{Diff}(M)$ with the property that the periodic points are dense in $\Omega(f)$. Then G is a Baire set.*

Unfortunately the C^r analogues for $r > 1$ of (6.8) and (6.9) are yet unproved.[22] Furthermore, in my opinion, it is important to find "conceptual" proofs of Pugh's important results.

We end by stating the three basic problems raised here.

(6.10) PROBLEMS. (a) Approximation problem:[23] For compact M, approximate (C^r, large r preferably) any $f \in \text{Diff}(M)$ by f' satisfying Axiom A and Axiom B. In this perhaps the most important property is Axiom A(a).

(b) Find all (in some sense) possible indecomposable hyperbolic sets of nonwandering points up to topological conjugacy.[24] This includes, as a special case, find all Anosov diffeomorphisms (such that $\Omega = M$).

(c) Find the possible ways of fitting the $W^s(\Omega_i)$ together to define $f: M \to M$ as in (6.3) and (6.5). This is sort of a generalized Morse theory type problem and essentially problem (6.6)a.[25]

It could happen that (6.10)(a) has a negative answer. This would add difficulty to the conjugacy problem! One would proceed by adding the corresponding counterexamples to those of this paper and enlarge the unifying framework.

On the other hand an affirmative answer to (6.10)(a) would imply that this survey gives the basic framework to the conjugacy problem and that answering (6.10)(b), (c) would be filling in the body.

I.7. Our goal in this section is to give at least a full sketch of proofs of the spectral decomposition theorem (6.2) and its corollary. In doing so we state a general stable manifold theorem and use it to show the existence of canonical coordinates on our hyperbolic sets. We first give some preliminary lemmas.

(7.1) LEMMA. (a) *If p is a periodic point of $f \in \mathrm{Diff}(M)$, and U is an open set in M such that $U \cap W^s(p) \neq \varnothing$, then the closure of $\bigcup_{m>0} f^m(U)$ $\supset W^u(p)$.*

(b) *Furthermore, if q is a second periodic point and $W^u(p) \cap W^s(q)$ contains a point of transversal intersection, then $\bigcup_{m>0} f^m(U) \cap W^s(q) \neq \varnothing$.*

PROOF. Note first that by replacing f by a power of f, we may as well assume p and q are fixed points to begin with. Since $f: W^u(p) \to W^u(p)$ is an expansion, it follows that if a neighborhood of p in $W^u(p)$ is in the closure of $\bigcup_{m>0} f^m(U)$, then so is all of $W^u(p)$. Thus we see that (a) is transformed into a local problem about a neighborhood N of p by replacing U by $f^k(U) \cap N$ for some large n.

In case f is linear in some chart about p the conclusion of (a) is easily checked directly, and finally the general case can be reduced to this one by an appeal to Hartman's (and Grobman's [74]) theorem [39], which gives a local topological equivalence to the linear case.

The second part of (7.1) can be proved with little trouble by using the linear Lemma 5.2 of [115] or one can use again Hartman's theorem and a topological intersection argument. The reduction to the local case is again clear. A stronger lemma than this, the "λ-lemma" is in [82].

(7.2) LEMMA. *Let $f: M \to M$ be a diffeomorphism with hyperbolic periodic points p_i, $i = 0, \cdots, n$ such that $p_0 = p_n$. Suppose for each $i = 0, \cdots, n-1$, $x_i \in W^u(p_i) \cap W^s(p_{i+1})$ is a point of transversal intersection. Then each x_i is nonwandering.*

PROOF. Let x_i for some i be as in (7.2) and U be a neighborhood of x_i in M. Then $(\bigcup_{m>0} f^m(U)) \cap W^s(p_j) \neq \varnothing$ every j using (7.1) inductively. By (7.1)(a) Closure $\bigcup_{m>0} f^m(U) \supset W^u(p_i)$. This shows x_i is nonwandering.

We next come to the general stable manifold theory which we put into the following form.

(7.3) GENERALIZED STABLE MANIFOLD THEOREM.[26] *Suppose $\Lambda \subset M$ is a hyperbolic set of $f \in \mathrm{Diff}(M)$ (that is, Λ is compact, invariant with the usual splitting of $T_\Lambda(M)$, see §I.6) with some metric on M. Then for each $x \in \Lambda$, there is an injective immersion $J_x^s = J_x : W^s(x) \to M$ with the following properties*:

(a) $x \in J_x(W^s(x))$, *and* $y \in J_x(W^s(x))$ *if and only if* $d(f^m(x), f^m(y)) \to 0$ *as* $m \to \infty$.

(b) $f(J_x(W^s(x)) = JK_{f(x)} W^s(f(x))$. *Let* $J_x(W^s(x)) = W^s(x)$ *now*.

(c) $\bigcup_{x \in \Lambda}(W^s(x)) = \{ y \in M \mid f^m(y) \to \Lambda,\ m \to \infty \}$.

(d) *For* $x, y \in \Lambda$, $W^s(x)$ *and* $W^s(y)$ *either coincide or are disjoint*.

(e) *The tangent space of* $W^s(x)$ *at* y *is* E_y^s *for each* $y \in \Lambda$ *(here* E_y^s *is part of the data of the hyperbolic splitting)*.

(f) $W^s(x)$ *and* $W^s(y)$ *are* C^1 *close on compact sets for* $x, y \in \Lambda$ *close*.

For Λ a point this is the stable manifold theorem for a fixed point, §I.2. In (7.3), $W^s(x)$ for $x \in \Lambda$ is called the stable manifold of x. The unstable manifold $W^u(x)$ is defined as the stable manifold of f^{-1} at x.

We will try now to give the history and background of (7.3). Of course it all starts with Λ a point from Poincaré, Perron, etc. as in §I.2. Anosov, using the basic work of Perron, proved (7.3) in the case Λ is all of M. This is the way he proved the structural stability in §I.3. Seeing the need for a more general version of stable manifold theory, because of Axiom A, I asked I. Kupka if he could give such a proof. In substance at least, he proved the above (7.3). All of the proofs in stable manifold theory, however, have been unsatisfactory from a conceptual point of view. On the other hand, at this writing it appears that the situation has been remedied by M. Hirsch. He seems to have a fully satisfactory proof of the above (7.3).

Added in proof. C. Pugh has a good proof of (7.3).

From the stable manifold theory we now construct what we call canonical coordinates on $\Omega(f)$ where $f \in \mathrm{Diff}(M)$ satisfies Axiom A. If $W^s(x)$ is as in (7.3), then we will denote an ϵ neighborhood of x in $W^s(x)$ in the intrinsic (metric) topology by $W^s(x, \epsilon)$. Then let $\tilde{W}^s(x, \epsilon)$ be the set $W^s(x, \epsilon) \cap \Omega$ etc.

(7.4) THEOREM (EXISTENCE OF CANONICAL COORDINATES).[27] *Suppose $f \in \mathrm{Diff}(M)$ satisfies Axiom A and that $x \in \Omega = \Omega(f)$. Then there is $\epsilon > 0$, independent of x, and a canonical map $I_x : V \to M$ where V is a neighborhood of $x \times x$ in $\tilde{W}^u(x) \times \tilde{W}^s(x)$, which is a homeomorphism of V onto a neighborhood of x in Ω. On $\tilde{W}^u(x, \epsilon) \times x$, I_x is the inclusion J_x^u and on $x \times \tilde{W}^s(x, \epsilon)$, I_x is the inclusion J_x^s. The map I_x is defined at $(p, q) \in V \subset \tilde{W}^u(x, \epsilon) \times \tilde{W}^s(x, \epsilon)$ as the unique intersection of $W^s(p, \epsilon)$ and $W^u(q, \epsilon)$ in M.*

PROOF. This follows directly from a systematic application of (7.3) and (7.2). The map is well defined into M by (7.3). The image of I_x is in Ω by (7.2) and the fact that the periodic points are dense in Ω. In a similar way, one checks that a neighborhood of x in Ω is in the image of I_x. The injectivity of I_x is a consequence of the stable manifold theorem, that the $W^s(p)$ for different p, either coincide or are disjoint.

Moving toward the proof of (6.2) we give first the following lemma:

(7.5) LEMMA. *Suppose* $f \in \mathrm{Diff}(M)$ *satisfies Axiom* A, $\Omega = \Omega(f)$ *the nonwandering points. Given* $x \in \Omega$, *suppose* N *is a neighborhood of* x *in* Ω *with the local product structure of* (7.4). *If* U *is any nonempty open subset of* N, *then* $\bigcup_{m \geq 0} f^m U$ *and* $\bigcup_{m \leq 0} f^m(U)$ *each contain a dense subset of* N.

PROOF. It is sufficient to consider just one of the two cases. Let q be a periodic point of U with stable and unstable manifolds $W^s(q)$ and $W^u(q)$. There exist such q since the periodic points are dense in Ω by Axiom A. Now let p be an arbitrary periodic point of N. There are points of transversal intersection $x \in W^u(p) \cap W^s(q)$, $x' \in W^u(q) \cap W^s(p)$, with x, x' in Ω. Then $x \in \bigcup_{m \leq 0} f^m U$, and so p is in the closure of $\bigcup_{m \leq 0} f^m U$. Since the periodic points are dense in N, this proves (7.5).

We now prove (6.2).

For $x \in \Omega$, let $N = N(x)$ be given as in the previous lemma and define $\Omega_x = \mathrm{Closure} \bigcup_{m \in Z} f^m(N)$. From the previous lemma it follows that Ω_x does not depend on any choices. In fact, it follows equally well from (7.5) that for x, $y \in \Omega$, either Ω_x and Ω_y coincide or are disjoint. Furthermore the union $\Omega = \bigcup_x \Omega_x$ is actually a finite union and all of the properties of (6.2) are checked very easily now using the previous lemma. Note that one obtains directly that any open set in Ω_i has a dense orbit (i.e., $\bigcup_m f^m(U)$ is dense). From Birkhoff [15] one then obtains topological transitivity.

We show how (6.3) follows from (6.2).

For each $x \in M$, $m \to \infty : f^m x \to \Omega$ from the definition of nonwandering. Given $x \in M$, we claim there is a unique i such that $f^m x \to \Omega_i$ as $m \to \infty$. For each $i = 1, \cdots, k$, choose open sets V_i, U_i such that $V_i \supset U_i \supset \Omega_i$, V_i disjoint and $f^{-1} U_i \cup f U_i \subset V_i$. Now, given $x \in M$, suppose there exist k, l such that $\Omega_k \cap \lim_{n \to \infty} f^n x \neq \varnothing$, $\Omega_l \cap \lim_{n \to \infty} f^n x \neq \varnothing$, $k \neq l$. Then there exist for each $j = 1, 2, \cdots$ positive integers m_j, l_j such that $f^{m_j} x \in U_k$, $f^{m_j + l_j} \in U_l$ where $m_j < m_j + l_j < m_{j+1}$ for each j. Then there exists n_j, $m_j < n_j < m_j + l_j$ such that $f^{n_j} x \notin \bigcup_{i=1}^{k} U_i$, for every j. Therefore $\lim_{j \to \infty} |f^{n_j} x \notin \Omega$, which is impossible. Thus $k = l$ and our assertion is proved, which in turn yields (6.3).

Note finally that the following proposition is clear from the previous material in this section.

(7.6) PROPOSITION. *Let Ω_i be as in (6.2).* (a) *Then for any $x \in \Omega_i$, $W^s(x)$ and $W^u(x)$ each contain a dense set of Ω_i.*

(b) *In particular if $f: M \to M$ is an Anosov diffeomorphism of a compact manifold with $\Omega = M$, then every stable manifold is dense in M.*

(c) $W^u(\Omega_i) \cap W^s(\Omega_i) = \Omega_i$.

One obtains (a) from the topological transitivity and the local product structure on Ω_i. (b) follows from (a). One checks (c) by first showing that $W^u(\Omega_i) \cap W^s(\Omega_i) \subset \Omega$ using (7.2) and the fact that the periodic points are dense in Ω. Then apply (6.2).

I.8. The goal of this section is to sketch the proofs of Theorems (6.5) and (6.6). We begin by introducing the generally useful notion of a filtration of a diffeomorphism.

A *filtration* then of $f \in \text{Diff}(M)$, M compact, is a sequence of closed submanifolds, $M = M_0 \supset M_1 \supset M_2 \supset \cdots \supset M_k = \varnothing$ where each M is an open subset together with its smooth boundary and $f(M_i) \subset$ interior of M_i.

(8.1) PROPOSITION.[28] *If $\{M_i\}$ is a filtration for $f \in \text{Diff}(M)$, then it is also a filtration for a C^0 approximation of f. Furthermore $\Omega(f) \cap \partial M_i = \varnothing$ for each i.*

This is easily checked. If $\{M_i\}$ is a filtration for $f \in \text{Diff}(M)$ then we can decompose $\Omega(f) = \Omega_1 \cup \cdots \cup \Omega_{k-1}$, Ω_i compact invariant, by defining $\Omega_i = \Omega \cap (M_i - M_{i-1})$. We will call this the Ω-decomposition of the filtration.

We assume now that f satisfies Axioms A and B, with Ω_i, $W^s(\Omega_i)$, etc. as in §I.6.

(8.2) PROPOSITION. *If $W^u(\Omega_i) \cap W^s(\Omega_j) \neq \varnothing$, then for any $p \in \Omega_i$, $q \in \Omega_j$, $W^u(p)$ and $W^s(q)$ have a point of transversal intersection.*

PROOF. This is a consequence of Axiom B and §I.7. Then by (7.1) we obtain

(8.3) COROLLARY. *If $W^u(\Omega_i) \cap W^s(\Omega_j) \neq \varnothing$, then $W^u(\Omega_j) \subset$ Closure $W^u(\Omega_i)$.*

(8.4) PROPOSITION. *If $W^u(\Omega_i) \cap W^s(\Omega_{i+1}) \neq \varnothing$, $i = 0, \cdots, m-1$ and $\Omega_0 = \Omega_m$, then all the Ω_i coincide.*

PROOF. Let periodic points $p_i \in \Omega_i$ for each i. Then by (8.2), $W^u(p_i)$ and $W^s(p_{i+1})$ have a point of transversal intersection q_i for each i. Apply (7.2) to see that each $q_i \in \Omega$. But then $q_i \in \Omega_i \cap \Omega_{i+1}$, so indeed the Ω_i coincide.

From (8.3) and (8.4) follows

(8.5) PROPOSITION. *The relation $\leq$ defined in §I.6 is a partial ordering.*

REMARK. One could define $\leq$ in these alternate ways as long as Axiom B was modified accordingly and the whole theory would be the same.

Alternate way 1. $\Omega_i \leq \Omega_j$ if Clos $W^u(\Omega_j) \cap$ Clos $W^s(\Omega_i) \neq \varnothing$.

Alternate way 2. $\Omega_i \leq \Omega_j$ if for any pair of neighborhoods U_i of Ω_i, U_j of Ω_j there is $x \in U_j$, $m \in Z^+$ such that $f^m(x) \in U_i$.

(8.6) PROPOSITION. *Suppose $f \in \text{Diff}(M)$ satisfies Axioms A and B, with Ω_i as in the spectral decomposition theorem. Then there exists a filtration of M, $M = M_0 \supset M_1 \supset \cdots \supset M_k = \varnothing$ where $\Omega \cap M_j - M_{j-1}$ is precisely one of the Ω_j (re-indexing if necessary, so $\Omega_j = \Omega \cap M_j - M_{j-1}$) of the spectral decomposition. Similarly one has a filtration for f^{-1}, $M = M_0' \supset M_1' \supset \cdots \supset M_k' = \varnothing$. Given neighborhoods U_i of Ω_i, one may choose the filtrations so that U_i contains $(M_i - M_{i-1}) \cap (M_i' - M_{i-1}')$ for each i. Consequently perturbations f' of f will satisfy $\Omega(f') \subset \cup U_i$.*

PROOF OF (8.6). One uses the diagram. Start by taking $M_k = \varnothing$, M_{k-1} a neighborhood of an extreme vertex (attractor), $M_{k-2} = M_{k-1} \cup V$ where V is a contracting neighborhood of a second attractor, continuing until all the attractors are used up, obtaining $M_l \supset M_{l-1} \supset \cdots \supset M_k$ say. Choose Ω_j so that Ω_j is next to only attractors in the diagram. One defines M_{l-1} as $M_l \cup W$ where W is of the form $\cup_{m_0 > m > 0} f^m(W_0)$, m_0 some large integer and W_0 is a small neighborhood of Ω_j. One may think of W as a neighborhood of $W^u(\Omega_j)$. By continuing in this way one obtains the desired filtrations.

Conversations with M. Shub were very useful in the following.

To prove Ω-stability[29] for $f \in \text{Diff}(M)$ satisfying Axioms A and B, one generalizes the procedure of Moser in the Appendix written by Mather. Instead of the map A defined there one uses the map

$$B: \text{Diff}(M) \times C^0(\Lambda, M) \to C^0(\Lambda, M)$$

defined by $B(g, h) = ghf^{-1}$. Here $C^0(\Lambda, M)$ is the space of continuous maps of Λ into M with the uniform topology with $\Lambda = \Omega_i$ an indecomposable piece of $\Omega(f)$. $C^0(\Lambda, M)$ is a manifold and B has its second partial derivative continuous in both variables. A version of the implicit function theorem yields a continuous map

$$h: \Lambda \longrightarrow M \quad \text{such that}$$

$$
\begin{array}{ccc}
\Lambda & \xrightarrow{h} & M \\
\downarrow f & & \downarrow g \\
\Lambda & \xrightarrow{h} & M
\end{array}
\quad \text{commutes}
$$

(providing g is close enough to f in C^1). There remains to complete the proof of Ω-stability of f, two things. First, h is 1-1. Here the Moser argument does not work. The proof goes by

(8.7) PROPOSITION. $f\colon \Lambda \to \Lambda$ *is expansive. This means there is* $\epsilon > 0$ *such that for any* x, $y \in \Lambda$, $x \neq y$, *there is* $n \in Z$ *with* $d(f^n x, f^n y) > \epsilon$.

The proof of this proposition follows from the fact that Λ is a hyperbolic set for f. Then the fact that h makes the above diagram commute and is close to the identity leads to the injectivity of h.

The second point to check is that when h is defined on each Ω_i as above, $h(\Omega(f)) = \Omega(g)$. Since the periodic points are dense in $\Omega(f)$ it follows that $h(\Omega(f)) \subset \Omega(g)$. Furthermore, by (8.6) we may assume that $\Omega(g) \subset$ small neighborhoods of Ω_i. Thus the proof of Ω-stability is reduced to the study of what happens in a small neighborhood of Ω_i. The stable manifold analysis finally takes care of this last point.

To finish our program, we must show that if g has been chosen close enough to f, then g also satisfies Axioms A and B. All of this is a consequence of the above provided $\Omega(g) = h(\Omega(f))$ has a hyperbolic structure for g. This proceeds by showing bounded hyperbolic linear maps on Banach spaces are open using the spectral theory at the end of [96], and then using the values of sections of the Banach space splitting to reconstruct the vector bundle splitting.

I.9. On basic sets of diffeomorphisms. This section is devoted to the problem of finding all the possible Ω_i that could occur in the spectral decomposition theorem (6.2) for diffeomorphisms of compact manifolds satisfying Axiom A. In other words we discuss what is known about Problem (6.10)(b). Expanding on this define a *basic set* of $f \in \text{Diff}(M)$ to be one of the Ω_i of (6.2) where f satisfies Axiom A.

(9.1) PROBLEM. Find all basic sets up to topological conjugacy.[30] Do they always have a rational zeta function?[31] Are they all locally the product of a Cantor set and a manifold?[32] Can they be given some type of algebraic structure?

One can consider a possibly more general, but localized picture by considering $f\colon U \to M$ with U an open set of M, f a diffeomorphism onto its image with $\Lambda \subset U$ satisfying

(9.2) (a) Λ is compact and $f(\Lambda) = \Lambda$,
      (b) Λ is a hyperbolic set for f (see §I.6),
      (c) the periodic points of f are dense in Λ,
      (d) f is topologically transitive on Λ,
      (e) $\bigcap_{m \in Z} f^m(U) = \Lambda$.

Since the basic sets have neighborhoods U which satisfy (9.2) we may consider

(9.3) PROBLEM. Find all Λ satisfying (9.2).

There is a construction which allows one to replace $f: U \rightarrow M$ by $g: \tilde{U} \rightarrow \tilde{U}$ in (9.2). On $U \times Z$ say points (x, m) and (x', n) are equivalent if $x' = g^{n-m}(x)$. Then the quotient space $\tilde{U}$ is a manifold (not necessarily Hausdorff) and one has a diffeomorphism $g: \tilde{U} \rightarrow \tilde{U}$ induced by $(x, m) \rightarrow (x, m+1)$. Define $\tilde{\Lambda} \subset \tilde{U}$ as the image of $(\Lambda, 0)$ under the projection $\pi: U \times Z \rightarrow U$.

·We will say that Λ in (9.2) is an *attractor* if U can be chosen so that $\bigcap_{m>0} f^m(U) = \Lambda$. Then when Λ is the basic set of a diffeomorphism satisfying Axioms A and B, an attractor corresponds to a vertex lying at an extreme point of the diagram of f.

A special case of (9.1) and (9.3) is to find the attractors.[33] Note that no symbolic flow of §I.5 can be an attractor, but that every Anosov diffeomorphism with $\Omega = M$ is already an attractor.

We will give an outline of all the ways we know of constructing basic sets; then we will go into more detail. First consider these four groups of basic sets:

(9.4) (a) Group 0. These are characterized by dimension $\Lambda = 0$.

(b) Group A. This is Anosov case with $\Omega = M$.

(c) Group DE. These are derived from expanding maps and will be described subsequently.

(d) Group DA. These are derived from Anosov diffeomorphisms and will also be described subsequently.

Furthermore one may take finite products of any of these to obtain other basic sets (see §I.10).

Group 0 is discussed first. This includes the finite Λ (periodic orbits) and the shift automorphisms Λ_N of §I.5. It seems likely to me that every basic set in group 0 is topologically conjugate to some closed invariant subset of Λ_N. Call $\Lambda \subset \Lambda_N$ a subshift if Λ is closed, invariant, and the periodic points are dense in Λ. One can ask generally to what extent the subshifts occur as basic sets.[34]

The following construction may shed some light on the above imbedding problem. Suppose Λ is a basic set of dimension 0 relative to $f: U \rightarrow M$, $U_1 \cup \cdots \cup U_N$ a disjoint union of local product neighborhoods of (7.4) which cover Λ (such U_i can always be found). Then let $g: \Lambda_N \rightarrow \Lambda_N$ be the shift automorphism on the N-symbols $U_1, \cdots, U_N$ and define $\alpha: \Lambda \rightarrow \Lambda_N$ by $\alpha(x)(m) = U_i$ where $x \in \Lambda$, $m \in Z$, and $f^m(x) \in U_i$. Then it is easily checked that α is continuous and equivariant. Can the U_i be chosen so that α is injective?

The following is a nontrivial example of a subshift as a basic set. We describe it in the following figure as a diffeomorphism of a 2-disk into itself which can be extended to a diffeomorphism $S^2 \to S^2$ by adding an expanding disk to the original.

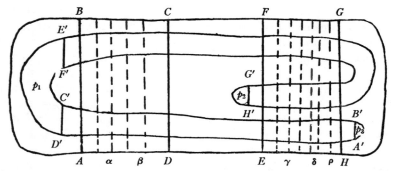

The construction follows in the pattern of those in §I.5. The diagram is simply given by ⦶ where the top vertex is an expanding fixed point and the bottom vertex corresponds to the periodic orbit consisting of p_1, p_2 and p_3. The reader can check that the middle vertex of the diagram corresponds to a subshift Λ of the shift on five symbols. In fact if Λ_5 is the shift space on the symbols $\alpha, \beta, \gamma, \delta, \rho$ corresponding to the indicated columns in the figure, then Λ consists of bi-infinite sequences which do not carry any of the following combinations $\beta\gamma$, $\beta\delta$, $\beta\rho$, $\alpha\alpha$, $\alpha\beta$, $\gamma\delta$, $\gamma\gamma$, $\gamma\rho$, $\delta\alpha$, $\delta\beta$, $\rho\alpha$, $\rho\beta$.

Generally speaking, relative to a shift automorphism a *block* is a finite sequence of symbols, e.g., β, etc. in the previous sentence. A subshift is said to be of *finite type* if it is of the form, all sequences which do not contain a certain finite set of blocks. Thus the above Λ is of finite type.

Related to the previous problems on basic sets of dimension 0 are the following theorems:

(9.5) THEOREM (O. LANFORD). *Every subshift of finite type has a rational zeta function.*

(9.6) THEOREM (R. BOWEN). *There exist subshifts with irrational zeta functions.*

In fact Lanford has improved Bowen's theorem to show that most subshifts have irrational zeta functions.

We don't go beyond the discussion of §I.3 on the Anosov case except to remark that it seems probable that if a basic set is a submanifold then the restriction of the diffeomorphism is conjugate to an Anosov diffeomorphism.

41

J. Moser has shown me an example of a basic set which is a sub-manifold but not a C^1 submanifold.

For the DE group we use the examples of Shub [108] of expanding endomorphisms of compact manifolds—see §I.10. For each expanding endomorphism, we will construct a basic set which is an attractor. This goes as follows.

Suppose then $f: M \rightarrow M$ is an expanding endomorphism of a compact manifold. Let D be the unit disk of dimension one larger than the dimension of M, with M imbedded in $D \times M$ as $0 \times M$. Let λ satisfy $0 < \lambda < 1$ and define $g_\lambda: D \times M \rightarrow D \times M$ by $g_\lambda(x, y) = (\lambda x, y)$. Next let $\phi: 0 \times M \rightarrow D \times M$ be a C^1 approximation of the map $0 \times M \rightarrow D \times M$, $(0, y) \rightarrow (0, f(y))$ such that ϕ is an embedding. This is possible by dimensional reasons (the Whitney imbedding theorem). Let T be a tubular neighborhood of $\phi(M)$ with fibers being the various components of $T \cap (D \times y)$, $y \in M$. Now extend ϕ to $\psi: D \times M \rightarrow T$ in a fiber preserving way so that ψ is even a diffeomorphism. Our desired map $D \times M \rightarrow D \times M$ is then the composition $\psi g_\lambda = h$ for λ small enough. It can be checked that for sufficiently small λ, the set $\Lambda = \bigcap_{m>0} h^m(D \times M)$ has a hyperbolic structure and is in fact a basic set. It is locally the product of a Cantor set and a manifold whose dimension is that of M.

The following figure gives $D \times M$ and its image under h when the starting point is the expanding endomorphism of $S^1 \rightarrow S^1$ defined by $z \rightarrow z^2$.[35]

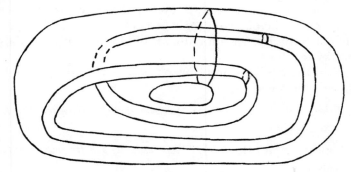

Finally we show how the DA group (9.4d) goes by giving the first case using an extended type of surgery on the Anosov diffeomorphism of the 2-torus.[36]

One changes the toral diffeomorphism on a small "square" neighborhood Q of the fixed point corresponding to $(0, 0)$ in R^2. Initially we have the square $Q = ABCD$ linearly mapped into $A'B'C'D'$ as in the following figure.

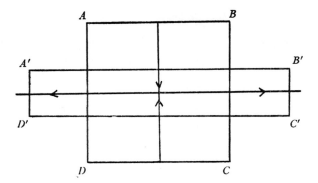

Without changing the diffeomorphism outside a neighborhood of the boundary of Q, we can change f on Q so that we have three fixed points in Q as illustrated in the following figure.

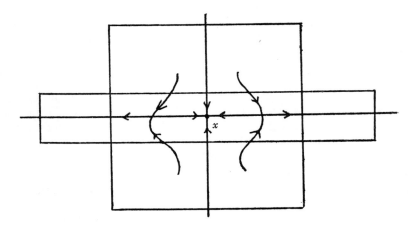

Now T^2 can be written as the union of a single two dimensional stable manifold of the fixed point x, $W^s(x)$ and a one dimensional basic Λ. We leave the (many) details to the reader.

One can apply this construction to any Anosov diffeomorphism.

As this was written, we received a very interesting manuscript of R. Williams [127] on 1-dimensional basic sets, which certainly appears to extend some of the above results.

I.10. **Final remarks on conjugacy problems.** We cover briefly a number of final miscellaneous points related to the diffeomorphism problem of part I. The first question is: what role do products play?

(10.1) PROPOSITION. *Let Ω_i be the set of nonwandering points of $f_i \in \mathrm{Diff}(M_i)$, $i = 1, 2$. Then the set of nonwandering points of the product $f_1 \times f_2 \in \mathrm{Diff}(M_1 \times M_2)$ is contained in $\Omega_1 \times \Omega_2$. Furthermore if the periodic points are dense in Ω_1 and Ω_2, then $\Omega = \Omega_1 \times \Omega_2$.*

This is checked easily from the definitions.

(10.2) PROPOSITION. *If $f_i \in \mathrm{Diff}(M_i)$, $i = 1$, 2 satisfying (2.2), then so does the product $f_1 \times f_2 \in \mathrm{Diff}(M_1 \times M_2)$.*

This follows from (10.1) and the fact that $W^s(p, q) = W^s(p) \times W^s(q)$. Furthermore one sees from the definitions that

(10.3) PROPOSITION. *The product of two Anosov diffeomorphisms is an Anosov diffeomorphism.*

The last two propositions are essentially contained in (equally easily checked)

(10.4) PROPOSITION. *If $f_i \in \mathrm{Diff}(M_i)$, $i = 1$, 2, both satisfy Axioms A and B, then so does the product $f_1 \times f_2$. Furthermore so does f_1^m, $m \in Z$, $m \neq 0$.*

Thus if $f_1 \in \mathrm{Diff}(M_1)$ is an Anosov diffeomorphism and if $f_2 \in \mathrm{Diff}(M_2)$ satisfies (2.2), then the product $f_1 \times f_2$ is Ω-stable (6.5). This product, however, is not structurally stable. Moreover, there is an open set U, in general, in $\mathrm{Diff}(M_1 \times M_2)$ near f with the property that U contains no structurally stable diffeomorphisms. This is described in [116]. It is the example mentioned in §I.1 to show that one had to weaken the concept of structural stability to get a successful theory. There is also an exposition of this fact in [11] and a further variant in [87].

We now define a modification of diffeomorphisms related to the notion of "surgery" in differential topology.

Suppose $f \in \mathrm{Diff}(M)$ has the property that there is a compact submanifold with boundary M_1, $\dim M = \dim M_1$, such that $f(M_1)$ is contained in the interior of M_1. Then it follows that $\Omega = \Omega(f)$ is equal to (interior $M_1) \cap \Omega \cup (M - M_1) \cap \Omega = \Omega_1 \cup \Omega_2$ where each Ω_i is compact and invariant. Furthermore, a similar decomposition can be done even for any sufficiently good C^0 approximation of f. This is a special case of the filtrations of §I.8.

For surgery, in addition to the above f suppose that $g: V \to V$ is a diffeomorphism of a compact manifold with boundary into its interior. Suppose further that there is a diffeomorphism $h: \mathrm{Cl}(M_1 - f(M_1)) \to \mathrm{Cl}(V - g(V))$ with $gh = hf$. An isotopy condition on f, g is sufficient to guarantee the existence of h at least in case Closure $(V-f(V))$ $=1 \times V$. Then one may replace M_1 by V and redefine f on M_1 by g on V to define $f': M' \to M'$, M' the modified manifold. This turns out to be a useful construction. One checks immediately that $\zeta_{f'}$ $= \zeta_f \circ \zeta_g \circ \zeta_{\bar{f}}^{-1}$ where $\bar{f}$ is f restricted to M_i.

A topological version of the ergodic theoretic concept, entropy, has been defined in [3]. In this paper, the authors showed that this topological entropy is positive for some of the examples we described in §§I.3 and 5. The following problem seems natural.

(10.5) PROBLEM. If Ω_i is a basic set of a diffeomorphism satisfying Axiom A (as described in the spectral decomposition theorem (6.2)), is the topological entropy of Ω_i positive?[37]

As J. Palis pointed out to me, any Anosov diffeomorphism will satisfy Axioms A and B. It is possible that applying some of the subsequent theorems, one could obtain an attack on the problem: For an Anosov diffeomorphism $f \in \mathrm{Diff}(M)$, M compact, must $\Omega(f) = M$? (See 3.4).

Up to now we have been investigating the dynamical system generated by a single $f \in \mathrm{Diff}(M)$. One can generalize this situation to a differentiable map (or endomorphism) $f: M \to M$ (without necessarily having an inverse). This f is not the generator of a group acting on M, but a semigroup Z^+ acting on M. M. Shub [108] has studied this problem and found that some of the previous results extend to cover this case and some new features are found here. We state some of these now.

The simplest new problem coming up in this context is the endomorphism of the complex numbers of absolute value one, $f: S^1 \to S^1$ defined by $f(z) = z^n$, $n \in Z$, $n > 1$. Is f structurally stable? As for diffeomorphisms, an endomorphism $f: M \to M$ is *structurally stable* if C^1 perturbations are conjugate to f by a homeomorphism. Shub gives an affirmative answer in the following more general proposition [108].

(10.6) PROPOSITION. *Suppose $f: S^1 \to S^1$ is C^1 with derivative everywhere > 1. Then f is conjugate by a homeomorphism to $z \to z^n$ where $n = degree f$.*

The general questions on endomorphisms of S^1 are not yet very well understood. On the other hand Shub has found a satisfactory generalization of (10.6) as follows.

Say that an endomorphism $f: M \to M$ of a complete Riemannian manifold is *expanding* if for each $v \in T_x(M)$, $\|Df^m(x)(v)\| \geq c\lambda^m \|v\|$, $m \in Z^+$, $c > 0$, $\lambda > 1$ independent of v, x. Examples of expanding endomorphisms are, of course, the circle map $z \to z^n$ as well as various products of these on tori.

(10.7) THEOREM (SHUB [108]). *Any two homotopic expanding endomorphisms of a compact Riemannian manifold are topologically conjugate.*

(10.8) COROLLARY. *Any expanding endomorphism of a compact Riemannian manifold is structurally stable.*

In view of (10.7), the following becomes a reasonable problem.

(10.9) PROBLEM. Find all expanding endomorphisms of manifolds[38] (up to conjugacy). Also, is (10.7) true for Anosov diffeomorphisms of compact manifolds?

Shub proves for expanding endomorphisms that the manifold is covered by Euclidean space, and has produced, besides those on tori, examples on the Klein bottle and nilmanifolds.

Presumably, eventually a systematic approach will include the Anosov diffeomorphisms and Shub expanding endomorphisms.[39] The unifying definition is, in fact, obvious.

For hyperbolic fixed points of an endomorphism $f: M \to M$, Shub defines stable and unstable manifolds, generalizing those for a diffeomorphism. In this case, however, W^s is no longer the image of a cell, but can be any manifold (i.e., map M into a point $p_0 \in M$ and take any small perturbation. Then the stable manifold of the fixed point will be M). On the other hand W^u is the image of a cell, but not under an immersion or a 1-1 map.

Shub generalizes the approximation theorem (6.7) to endomorphisms.

Previously, holomorphic endomorphisms of the Riemann sphere had been studied by G. Julia ([50], his "prize memoir").[40] Stein and Ulam [119] have made a study of certain polynomial endomorphisms of the plane using computing machines.

An extremely interesting problem is the study of maps of finite dimensional manifolds into Diff(M). What are generic properties of such maps? This is called bifurcation theory.[41] The most important work on this subject is that of J. Sotomayor [118]. He considers maps of an interval into the space of flows on 2-manifolds, and obtains a pretty complete picture in this case.

APPENDIX TO PART I: ANOSOV DIFFEOMORPHISMS

BY JOHN MATHER

In this Appendix, I give an exposition of Moser's proof that Anosov diffeomorphisms are structurally stable. (See Theorem 3.3 of §I.3.)

The main novelty in my presentation is the use of the language of manifolds of mappings, which seems to result in conceptual simplification. I would like to thank R. Abraham for suggesting that Moser's proof might be simply expressed in the language of manifolds of mappings.

We let M be a compact C^∞ manifold and $f: M \to M$ a C^1 diffeomorphism. We let D denote the topological space of diffeomorphisms of M into itself with the C^1 topology, H the topological space of homeomorphisms of M into itself with the C^0 topology (compact open topology), and C the C^∞ Banach manifold of continuous mappings of M into itself, where the topology is the C^0 topology and the manifold structure is defined in a manner similar to that by which the manifold structures on sets of mappings are defined in [1] (of Appendix).

THEOREM 1 (ANOSOV). *If f is an Anosov diffeomorphism, then f is structurally stable. More precisely, there exists a neighborhood U of the identity of M, id_M, in H, a neighborhood V of f in D and a continuous mapping $g \to h(g)$ of V into U such that for all $g \in V$, $h = h(g)$ is the unique solution in U of the equation*

$$hg = fh.$$

If E is a vector bundle over M, we let $\Gamma(E)$ denote the Banachable R-vector space of continuous sections of E over M, with the C^0 topology. If, further, $x \in M$, we let E_x denote the fiber of E over x. We let $f_*: \Gamma(TM) \to \Gamma(TM)$ be the continuous linear mapping given by $f_*(z) = Df \circ z \circ f^{-1}$. We will consider $\Gamma(TM)$ as a Banach space, with any norm which induces its topology.

LEMMA 1. *If f is an Anosov diffeomorphism, then $f_* - \mathrm{id}$ is an isomorphism.*

REMARK. The converse is also true, but will not be proved here.

PROOF. By the hypothesis, there exists a splitting of $T(M)$ into a continuous Whitney sum $T(M) = E^s + E^u$, invariant under Df, such that $Df: E^s \to E^s$ is contracting and $Df: E^u \to E^u$ is expanding. Let $f_0 = f_* | \Gamma(E^s)$ and $f_\infty = f_* | \Gamma(E^u)$. Then there exists $C > 0$, $0 < \lambda < 1$ such that for all $m \in Z^+$,

$$\|f_0^m\| < C\lambda^m, \qquad \|f_\infty^{-m}\| < C\lambda^m.$$

It follows that $f_0 - \mathrm{id}$ and $f_\infty^{-1} - \mathrm{id}$ are automorphisms of $\Gamma(E^s)$ and $\Gamma(E^u)$, resp. In fact,

$$(f_0 - \text{id})^{-1} = -\sum_{j=0}^{\infty} f_0^j,$$

$$(f_\infty^{-1} - \text{id})^{-1} = -\sum_{=0}^{\infty} f_\infty^{-j}.$$

Hence, $f_\infty - \text{id} = -f_\infty(f_\infty^{-1} - \text{id})$ is an automorphism. The lemma follows immediately.

Let $A: D \times C \times D \to C \times C$ be given by $A(g_1, h, g_2) = \langle h \circ g_1, g_2 \circ h \rangle$.

LEMMA 2. *A is once differentiable in its second variable and its "partial derivative,"*

$$D_2 A: \quad D \times TC \times D \to TC \times TC$$

is continuous in all variables. Moreover $p \circ L$ is an isomorphism, where

$$L = D_2 A \mid (f \times (TC)_{\text{id}} \times f): \quad (TC)_{\text{id}} \to (TC)_f \times (TC)_f$$

and p denotes the projection of $(TC)_f \times (TC)_f$ on $(TC)_f \times (TC)_f / \text{diagonal}$.

PROOF. The first sentence follows from the methods of [1]. Also by the methods of [1], we may make the identifications $((TC)_{\text{id}} = \Gamma(TM)$ and $(TC)_f = \Gamma(f^*TM)$. With respect to these identifications L is given by

$$z \to \langle z \circ f, Df \circ z \rangle.$$

Let $S: (\Gamma(f^*TM) \times \Gamma(f^*TM))/\text{diagonal} \to \Gamma(TM)$ be the isomorphism induced by $\langle s, t \rangle \to t \circ f^{-1} - s \circ f^{-1}$. Then

$$S \circ p \circ L = f_* - \text{id}.$$

Hence, the second sentence follows from Lemma 1, completing the proof of Lemma 2.

By Lemma 2 and a suitable version of the implicit function theorem, there exists a neighborhood V_1 of f in D and a neighborhood U_1 of id_M in C such that for all $g_1, g_2 \in V$ there exists a unique $h = u(g_1, g_2) \in U_1$ such that

$$A(g_1, h, g_2) \in \text{diagonal}$$

i.e.,

(1) $h \circ g_1 = g_2 \circ h$

and such that $\langle g_1, g_2 \rangle \to u(g_1, g_2)$ is continuous.

Let U_2 be a neighborhood of id_M in U_1 such that for all $h_1, h_2 \in U_2$, $h_1 \circ h_2 \in U_1$ and let V be a neighborhood of f in V_1 such that for all g_1, $g_2 \in V$, $u(g_1, g_2) \in U_2$. For all $g \in V$, set $h(g) = u(g, f)$ and $h^-(g) = u(f, g)$.

Setting $h=h(g)$, $h^-=h^-(g)$, we have $hg=fh$ and $h^-f=gh^-$. Hence $h^-hg=h^-fh=gh^-h$ and $hh^-f=hgh^-=fhh^-$. Since (1) has a *unique* solution $h\in U_1$ for $g_1, g_2\in V_1$, it follows that $hh^-=h^-h=\text{id}_M$. Hence h is a homeomorphism, so the theorem follows with $U=U_2\cap H$.

REFERENCE

1. R. Abraham, *Lectures of Smale on differential topology*, Lecture notes, Columbia University, New York, 1962.

PART II. FLOWS

II.1. Introduction to flows. We shift now our survey to the case of the group $G=R$, the real numbers, acting on a manifold M, which for simplicity we will assume compact most of the time. Thus we are studying a 1-parameter group of diffeomorphisms, ϕ: $R\rightarrow\text{Diff}(M)$ with ϕ_0: $M\rightarrow M$ the identity and $\phi_t\phi_s=\phi_{t+s}$. We will call this set of data, or ϕ, simply a *flow*. A flow ϕ_t: $M\rightarrow M$ defines (or generates) a tangent vector field on M; i.e., for each $x\in M$ define $X(x)\in T_x(M)$ by

$$(1.1) \qquad d\phi_t(x)/dt]_{t=0} = X(x).$$

Thus $X(x)$ is a tangent vector at x on M and $\phi_t(x)$ is the solution of the ordinary differential equation (1.1) with initial condition $\phi_0(x)=x$. Then the *orbit* (of ϕ) through x, $t\rightarrow\phi_t(x)$ coincides with the solution of the first order, autonomous (i.e., $X(x)$ doesn't depend on t), ordinary differential equation (1.1).

Conversely, given an ordinary differential equation, simple methods reduce it to the first order autonomous case and thus one obtains the situation in (1.1) with $X(x)$ given. The fundamental existence theorems of ordinary differential equations (see [25], [56]) yield a solution $\phi_t(x)$ such that $\phi_0(x)=x$, at least locally, i.e., for $|t|<\epsilon$. Furthermore these local solutions may be pieced together (see [56]) and frequently this leads to a flow on M. Certainly if M is compact, every (smooth of course) tangent vector field defines a unique flow in this way. In the noncompact case one may change the vector field by a scalar factor to obtain one which defines a global flow. We will consider here only the case where ϕ_t: $M\rightarrow M$ is defined for all t, or an action of R on M, i.e., a flow.

Most of Part II is the carrying over of Part I to this slightly more complicated case. We will emphasize some of the special features and new interesting problems encountered in this 1-parameter case.

There are three possible types of orbits of a flow ϕ_t: $M\rightarrow M$. First

49

x is *fixed point* of the flow if $\phi_t(x) = x$ all $t \in R$. A fixed point x can also be characterized as a zero of the vector field defined by ϕ. Secondly a *closed orbit* of ϕ_t: $M \to M$ is the orbit through some x with $\phi_t(x) = x$ some $t \neq 0$. Usually a closed orbit is taken to mean exclusion of the fixed point case so there is a minimum *period* $t_0 > 0$ such that $\phi_{t_0}(x) = x$. Finally if $t \to \phi_t(x)$ is injective, then the orbit through x is not one of the above types and could be called an ordinary orbit.

In topologizing actions of R we assume M is compact. Then the flows as we saw above correspond precisely to tangent vector fields on M. The C^r, $r > 0$, vector fields on M form a linear space and with a C^r norm, $r < \infty$, a Banach space which we denote by $\chi(M)$. A *generic property* of flows will be a property true for a Baire set in $\chi(M)$. The most obvious generic property is that the set of zeros of $X \in \chi(M)$ is finite [114].

Proceeding as in §I.1 we look for a suitable equivalence between two flows ϕ_t and ψ_t on M. A *conjugacy* between ϕ_t and ψ_t is a homeomorphism h: $M \to M$ such that $h\phi_t(x) = \psi_t(hx)$. Such an equivalence relation preserves the minimum period of a periodic orbit and thus a conjugacy class will not in general be invariant under perturbation. This implies the need of a weaker notion of equivalence. We say that flows ϕ_t, ψ_t are *topologically equivalent* if there is a homeomorphism of M sending orbits of ϕ_t into orbits of ψ_t. If perturbations in $\chi(M)$ do not change the topological equivalence class of $X \in \chi(M)$, then X is called *structurally stable*. This concept was introduced in 1937 [6] by Andronov and Pontrjagin for ordinary differential equations on the 2-dimensional disk. On compact 2-dimensional manifolds, the structurally stable flows form a dense open set, simply characterized (Peixoto [84], see also §II.2). However in every dimension higher than three there exist compact manifolds on which the structurally stable flows are not dense [116] (see also [87]). Parallel to Part I this leads to a weakening of topological equivalence as follows.

For a flow ϕ_t on M, G. D. Birkhoff [15] has defined $x \in M$ to be a wandering point if there is some neighborhood U of x in M with $(\bigcup_{|t| > t_0} \phi_t(U)) \cap U = \varnothing$ for some $t_0 > 0$.

The nonwandering points (those which are not wandering) form a closed invariant subset of M denoted by $\Omega = \Omega(\phi_t)$. We will say that flows ϕ_t, ψ_t are *topologically equivalent on* Ω if there is an orbit preserving homeomorphism h: $\Omega(\phi_t) \to \Omega(\psi_t)$. Then ϕ_t is Ω-*stable* if sufficiently small perturbations (measured in terms of the corresponding $X \in \chi(M)$ of course) are topologically equivalent on Ω to ϕ_t. It is an important problem to discover whether Ω-stable flows are dense in $\chi(M)$.

We end this section by giving a direct relation between the flows discussed here and the diffeomorphism questions of Part I, [114].

A compact submanifold Σ of codimension one of a compact manifold M is called a *cross-section* for a flow ϕ_t on M if Σ intersects every orbit, has transversal intersection with the flow and whenever $x \in \Sigma$, $\phi_t(x) \in \Sigma$ for some $t > 0$. Then ϕ_t induces a diffeomorphism $f: \Sigma \to \Sigma$ by $f(x) = \phi_{t_0}(x)$ where t_0 is the first $t > 0$ with $\phi_t(x) \in \Sigma$ (see [114] for more details). The topological equivalence class of ϕ_t is determined by the topological conjugacy class of f. Orbits of f are in a natural 1-1 correspondence with those of ϕ_t by $\{f^m(x) \mid m \in Z\} \to \{\phi_t(x) \mid t \in R\}$, each $x \in \Sigma$. Compact orbits are preserved under this correspondence; thus periodic points of f correspond to closed orbits of ϕ_t. There can be no fixed points of ϕ_t when there is a cross-section. Cross-sections were used by Poincaré and Birkhoff (see, e.g., [19]).[42]

There is a converse construction of some importance. Given a diffeomorphism f of a (compact) manifold Σ we will construct a flow, canonically, on a manifold M_0 of one dimension higher, called the *suspension* of f. This goes as follows. Let $\alpha: \Sigma \times R \to \Sigma \times R$ be defined by $\alpha(x, u) = (f(x), u+1)$. Then $\{\alpha^m\} = Z$ operates freely on $\Sigma \times R$ and the orbit space is a manifold M_0. Furthermore the flow $\phi_t: \Sigma \times R \to \Sigma \times R$ defined by $\psi_t(x, u) = (x, u+t)$ induces a flow ϕ_t on M_0 which is our suspension of f. Clearly M_0 will have a cross-section $\Sigma_0 = \pi(\Sigma \times 0) \subset M$ where $\pi: \Sigma \times R \to M_0$ is the quotient map. It is easy to check that the associated diffeomorphism of (ϕ_t, Σ_0), $f_0: \Sigma_0 \to \Sigma_0$ is differentiably conjugate to our initial $f: \Sigma \to \Sigma$. Furthermore if an arbitrary flow $\phi_t: M \to \Sigma$ has a cross-section $f: \Sigma \to \Sigma$ whose suspension is $\phi'_t: M_0 \to M_0$, then ϕ_t and ϕ'_t are equivalent by an orbit preserving homeomorphism.[43]

This notion of suspension is useful because it allows one immediately to transfer all the examples in Part I, i.e., the diffeomorphisms of §I.2, Anosov diffeomorphisms as well as those of §§I.5 and I.6, to examples of flows. From the above remarks, all the stability properties of the diffeomorphism examples are kept by the suspended flows.

II.2. **The simplest examples of Ω-stable flows.** We will say that a fixed point x of the flow $\phi_t: M \to M$ is *hyperbolic* if x is a hyperbolic fixed point of the diffeomorphism $\phi_1: M \to M$. An alternate way of saying this is as follows: If x is a fixed point of the flow $\phi_t: M \to M$, then the derivative $D\phi_t(x): T_x(M) \to T_x(M)$ defines a linear representation of the real line and so can be written in the form $D\phi_t(x) = e^{tA}$ where A is a linear endomorphism of $T_x(M)$. Then x is hyperbolic if and only if none of the eigenvalues of A have real part equal to zero.

(2.1) PROPOSITION. *If x is a hyperbolic fixed point for the flow ϕ_t, the stable manifold $W^s(x)$ of x relative to ϕ_1 is invariant under ϕ_t for every t and is contracting for every $t > 0$.*

Then $W^s(x)$ will be called the *stable manifold* of x for the flow ϕ_t. One may apply I.(2.1) to obtain properties of $W^s(x)$.

Now suppose that $x \in M$ is in a closed orbit γ of the flow $\phi_t : M \to M$. There is a submanifold V of codimension one passing through x and transversal to γ. Then V serves as a local version of the cross-section of §II.1, defining a local diffeomorphism $f : U \to V$, $f(x) = x$, where U is a neighborhood of x in V. We say that γ is a *hyperbolic closed orbit* of ϕ_t whenever x is a hyperbolic fixed point of f. It is easily checked that this definition is independent of the choices $x \in \gamma$ and V (see [114]). The local stable manifold $W^s_{\mathrm{loc}}(x, f)$ of x for f in U defines the *stable manifold* $W^s(\gamma)$ of γ by $W^s(\gamma) = \bigcup_{t \in R} \phi_t(W^s_{\mathrm{loc}}(x, f))$. Then $W^s(\gamma)$ is a 1-1 immersed cell bundle over S' (either a cylinder $R^k \times S^1$, or a generalized Möbius band). For more details see [114].

The *unstable manifolds* of hyperbolic fixed points and closed orbits of ϕ_t are defined as the stable manifolds of $\psi_t = \phi_{-t}$.

For the suspension of a toral diffeomorphism (§I.3), the closed orbits are hyperbolic and dense in M; but hyperbolic fixed points of any flow are necessarily isolated fixed points.

We now describe the analogue of the diffeomorphisms of §I.2 as flows $\phi_t : M \to M$, M compact, which satisfy

(2.2) (1) $\Omega(\phi_t)$ is the union of a finite number of fixed points $x_1, \cdots, x_m$ and a finite number of closed orbits $\gamma_1, \cdots, \gamma_n$ of ϕ_t.

(2) The x_i, γ_j are all hyperbolic.

(3) The stable manifolds and unstable manifolds of the x_i, β_j intersect each other only transversally.

(2.3) THEOREM [109].[44] *Suppose the flow $\phi_t : M \to M$ satisfies (2.2). Then (a) Each stable manifold W^s_k of the x_i and γ_j is imbedded and $M = \bigcup_{k=1}^{m+n} W^s_k$ (disjoint union).*

(b) *The closure of one W^s_k is the union of certain W^s_i. Let $W^s_i \leq W^s_k$ if W^s_i is in the closure of W^s_k. Then $\leq$ is a partial ordering. If $W^s_i \leq W^s_k$ then* $\dim W^s_i \leq \dim W^s_k$.

(c) *One has the following Morse inequalities:*

$$M_0 \geqq B_0,$$
$$M_1 - M_0 \geqq B_1 - B_0,$$

. . . .

$$\sum_{k=0}^{n} (-1)^k M_k = \sum_{k=0}^{n} (-1)^k B_k.$$

Here B_i is the ith betti number coefficients Z_2, and $M_i = a_i + b_i + b_{i+1}$ where a_i is the number of x_j with dimension $W^s(x_j) = i$ and b_i is the number of γ_j with dimension $W^s(\gamma_j) = i+1$.

One can see that II.(2.3) is quite analogous to I.(2.3). There are a couple of special features in the present situation however. For example

(2.4) THEOREM (PEIXOTO [84]).[45] *If* dim $M = 2$, *then the flow* ϕ_t *satisfies* (2.2) *if and only if it is structurally stable.*

In this case the corresponding $X \in \chi(M)$ form an open and dense set.

This theorem gives a rough but quite good picture of flows on compact 2-manifolds. It solves the first basic problem for 2-dimensional flows.

A gradient flow $\phi_t \colon M \to M$ on a compact Riemannian manifold is defined by a C^r function $f \colon M \to R$ in the following way. The derivative $Df(x)$ of f at x is a cotangent vector at x and the Riemannian metric converts this into a tangent vector $X(x) = (\text{grad } f)(x)$ at x. By the familiar procedure (§II.1) from $X(x)$ we obtain our gradient flow ϕ_t.

(2.5) THEOREM [109]. *The flows on any given* M *satisfying* (2.2) *contain an open and dense subset of all gradient flows.*

Since every manifold possesses Riemannian metrics, we see that from (2.5) every manifold exhibits flows satisfying (2.2). Recall the existence of the diffeomorphisms of §I.2 was obtained in this way. Theorem (2.5) gives the bridge between the usual Morse theory for functions on manifolds and the work in this section. This even brings the subject here close to handlebody theory in differential topology and Poincaré duality on a manifold (this is the duality between the stable and unstable manifolds of a gradient flow).

See [95] for one definitely nongradient type example satisfying (2.2). See also [63], [97] for related papers.

I have just received a manuscript of K. Meyer [15] in which

"energy functions" are constructed for the flows described in this section.

II.3. Anosov flows. Consider first 1-*parameter groups* of *vector space bundle automorphisms* $\phi_t\colon E \to E$. Here E is a vector space bundle and ϕ_t is a flow on E such that for each t, $\phi_t\colon E \to E$ is a bundle automorphism (i.e., linear on fibers). For example if $\psi_t\colon M \to M$ is a flow on a manifold, the derivatives at each t, $\phi_t = D\psi_t\colon T(M) \to T(M)$, define a 1-parameter group of vector space bundle automorphisms. Assuming E is a Riemannian vector space bundle, say that such a flow $\phi_t\colon E \to E$ is *contracting* if there are constants c, $\lambda > 0$ such that $\|\phi_t(v)\| \leq ce^{-\lambda t}$, all $v \in E$, $t > 0$.

Then ϕ_t is *expanding* if ϕ_{-t} is contracting and this is equivalent to the existence of $c_1 > 0$, $\mu > 0$ such that $\|\phi_t(v)\| \geq c_1 e^{\mu t}$ all $t > 0$, $v \in E$ (compare §I.3).

An Anosov flow on a complete Riemannian manifold M (or just a manifold in case M is compact) is a flow ϕ_t whose induced flow $D\phi_t\colon T(M) \to T(M)$ on the tangent bundle is hyperbolic in the following sense: The tangent bundle $T(M)$ can be written as the Whitney sum of 3 invariant subbundles, $T(M) = E_1 + E_2 + E_3$ where on $E^u = E_1$, ϕ_t is expanding, on $E^s = E_2$, ϕ_t is contracting and E_3 is the 1-dimensional bundle defined by differentiating ϕ_t with respect to t.

Examples of Anosov flows are obtained readily from §I.3 and the following easily proved proposition.

(3.1) PROPOSITION. *If $f\colon M \to M$ is an Anosov diffeomorphism of a compact manifold, then the suspension of f is an Anosov flow.*

Another important class of examples of Anosov flows are the geodesic flows on the tangent bundles of Riemannian manifolds of negative (possibly varying) curvature (see [8], [13]).

(3.2) THEOREM (ANOSOV [9]). *If $\phi_t\colon M \to M$ is an Anosov flow of a compact manifold it is structurally stable. Also if $\Omega = M$, the periodic orbits will be dense. Finally if there is an (Lebesgue) invariant measure, then ϕ_t is ergodic.*

Applied to the geodesic flows on the tangent bundles of manifolds M with negative curvature, (3.2) yields ergodicity, thus solving an old problem. The constant negative curvature case as well as the case of two dimensional M had been done earlier by G. Hedlund [42] and E. Hopf [45], [46]. See also [34] and [64].

Again as in §I.3. there is the very important problem of finding all Anosov flows on compact M (especially when $\Omega = M$). Progress on this problem might contribute to the problem of what manifolds can have Riemannian metrics of negative curvature. Is this class bigger than the class of manifolds which possess Riemannian metrics of constant negative curvature? On this point see the problem of Calabi in [51].

II.4. **On counting closed orbits.** For counting the fixed points (at least algebraically) of a diffeomorphism, the Lefschetz Trace Formula provides a satisfactory method (see §I.4). This also applies to periodic points, and for suspended flows, these methods will give us some answers as to the nature of closed orbits. For flows in general, it is an outstanding problem to find methods which will tell if there are closed orbits and how many.

Seifert's problem [105] is the best known question exemplifying this lack of knowledge. That is, does a flow on S^3 (continuous or differentiable) have a closed orbit or a fixed point?[46] A related question is: does X, a smooth vector field on $D^2 \times S^1$, the 2-disk cross the circle, transversal to the boundary, have a closed orbit or a singular point? Related to these questions are papers of Fuller [32], and A. Schwartz [104].

Thus an analogue of the zeta function for diffeomorphisms of §I.4 seems quite remote for flows. However we will mention a wild idea in this direction.

Let $\Gamma = \Gamma(\phi_t)$ be the set of closed orbits of the flow $\phi_t: M \to M$ where we will assume M to be compact and that there are no fixed points. For $\gamma \in \Gamma$, define $l(\gamma)$ to be the period (minimal period, that is) of γ, i.e., $l(\gamma)$ is the first $t > 0$ such that $\phi_t(x) = x$ for some $x \in \gamma$. We will assume then that the flow satisfies the generic property, $\{\gamma \in \Gamma \mid l(\gamma) \leq c\}$ is finite for each positive c (that this is generic follows from II.(5.6)).

Then define formally (another zeta function!) $Z(s)$ to be the infinite product

$$Z(s) = \prod_{\gamma \in \Gamma} \prod_{k=0}^{\infty} (1 - [\exp l(\gamma)]^{-s-k}).$$

The question is: does $Z(s)$ have nice properties for any general class of flows ϕ?[47] In this direction we consider the case that ϕ is the suspension of a diffeomorphism $f: V \to V$ where the zeta function (of Weil, Artin-Mazur, §I.4) is rational.

(4.1) THEOREM. *If the zeta function of $f: V \to V$ is rational, then $Z_\phi(s) = Z(s)$ where ϕ is the suspension of f converges in a half-plane to an analytic function of s, and has an analytic continuation to a meromorphic function. Furthermore the zeros and poles of this meromorphic function can be computed explicitly in terms of those of ζ_f.*

PROOF (LARGELY DUE TO NARASIMHAN). If $K_m = K_m(f)$, $m = 1, 2, 3, \cdots$ (as in §I.4) denotes the number of periodic points of f of minimum period m, we get directly from the definition of $Z(s)$

$$Z(s) = \prod_{m=1}^{\infty} \prod_{k=1}^{\infty} (1 - e^{-m(s+k)})^{K_m/m}.$$

Let

$$W(s) = \prod_{m=1}^{\infty} (1 - e^{-ms})^{K_m/m}.$$

Then

$$-\log W(s) = \sum_{m=1}^{\infty} \frac{1}{m} K_m \log \left(\frac{1}{1 - e^{-ms}} \right)$$

$$= \sum_{m,r \geq 1} \frac{1}{mr} K_m e^{-mrs}$$

$$= \sum_{n \geq 1} \frac{1}{n} e^{-ns} \sum_{m/n} K_m.$$

Assuming at first that the zeta function $\zeta(t)$ of f is of the form $\zeta(t) = (1 - \lambda t)^{-1}$, we have (I.(4.1), (4.8)) $\sum_{m/n} K_m = \lambda^n$. Thus $-\log W(s) = \sum (1/n)(\lambda/e^s)^n = -\log(1 - \lambda/e^s)$ or $W(s) = 1 - \lambda/e^s$ and

$$Z(s) = \prod_{k=0}^{\infty} W(s+k) = \prod_{k=0}^{\infty} (1 - \lambda/e^{s+k}).$$

Then we can see that $Z(s)$ is entire because it is the uniform limit, in every compact set, of entire functions. Incidentally one sees from the explicit form of $Z(s)$, a functional equation $Z(s+1) = Z(s)e^s/(e^s - \lambda)$. Finally the zeros are clearly the solutions of $e^{s+\mu} = \lambda$, $k = 0, 1, 2, \cdots$ or $s + k = \log \lambda + 2\pi in$, $n \in Z$.

In the general case we have $\zeta(t) = \prod_{i,j}(1 - \mu_j t)/(1 - \lambda_i t)$ and $\sum_{m/n} K_m = \sum_i \lambda_i^n - \sum_j \mu_j^n$. Thus we obtain $-\log W(s) = -\log \prod_{i,j}(1 - \lambda_i/e^s)(1 - \mu_j/e^s)^{-1}$, so $Z(s) = \prod_{i,j} \prod_{k=0}^{\infty}(1 - \lambda_i/e^{s+k})$ $\cdot (1 - \mu_j/e^{s+k})^{-1}$. The zeros are of the form $s = \log \lambda_i + 2\pi ni - k$ and the poles $s = \log \mu_j + 2\pi ni - k$ (distinguish the i's!). This proves (4.1).

The following question then arises. Suppose $\phi_t \colon M \to M$ is the suspension of f as in (4.1) with ζ_f rational and f satisfying Axioms A and B of §I.6. Suppose even that M is the 2-dimensional toral diffeomorphism. Now let $\psi_t \colon M \to M$ be close to ϕ_t. Does $Z_{\psi_t}(s)$ have a meromorphic continuation to all of C? An affirmative answer would be roughly necessary and sufficient condition for $Z(s)$ to be useful. I must admit a positive answer would be a little shocking! A way of looking at this problem is the following. The canonical cross-section Σ for ϕ_t is also a cross-section for ψ_t and the time of first return for ψ_t is defined by a smooth function $\rho: \Sigma \to R^+$ (R^+ the positive reals) which will be close to the constant function 1. There is a natural 1-1 correspondence $\gamma \to \gamma'$, $\Gamma \to \Gamma'$ from the set of closed orbits of ϕ_t to those of ψ_t using Ω-stability.

Let $\lambda_\gamma = l(\gamma')$, so $\lambda_\gamma = \sum_{x_i \in \gamma' \cap \Sigma} \rho(x_i)$ and $-\log W_1(s) = \sum_{\gamma, r} (1/r) e^{-\lambda_\gamma r s}$ where $W_1(s)$ corresponds to the $W(s)$ of the previous proof. Is there sufficient regularity in the λ_γ to continue W_1 meromorphically?

There are two other remarks we wish to make about $Z(s)$. First if ϕ_t is the geodesic flow for a 2-manifold of constant negative curvature, then $Z(s)$ is meromorphic. In this case it is precisely the Selberg zeta function [106], which Selberg defined directly in terms of SL(2, R) and a certain uniform discrete subgroup Γ. Selberg proved that it is meromorphic in this case and found its zeros and poles as well. Sinai and Langlands pointed out to me this interpretation of the Selberg zeta function and this motivated my using it here.

Finally we pose the question, how generally do flows have the $l(\gamma)$ growing slowly enough so that $Z(s)$ has a half plane of convergence?[48]

II.5. **Spectral decomposition of flows.** One can extend Axioms A and B of §I.6 to flows. This goes as follows. For flows ϕ_t on compact manifolds M, we have

(5.1) AXIOM A'. *The fixed points of ϕ_t are each hyperbolic. The non-wandering points Ω consist of this finite set of fixed points F and the closure Λ of the closed orbits; Λ and F are disjoint. Finally the derived flow restricted to the tangent bundle restricted to Λ, $D\phi_t: T_\Lambda(M) \to T_\Lambda(M)$ is hyperbolic (defined analogously to the Anosov flow in §II.3).*

Topologically transitive for a flow again means that there is a dense orbit.

(5.2) THEOREM (SPECTRAL DECOMPOSITION). *If $\phi_t: M \to M$ satisfies Axiom A', then Ω can be written uniquely as the disjoint union $\Omega_1 \cup \Omega_2 \cup \cdots \cup \Omega_k$ where each Ω_i is closed, invariant and each $\phi_t: \Omega_i \to \Omega_i$ is topologically transitive.*

(5.3) COROLLARY. $M = \bigcup_{i=1}^k W^s(\Omega_i)$ *(disjoint union, canonically) where each* $W^s(\Omega_i) = \{x \in M \mid \phi_t(x) \to \Omega_i\}$.

(5.4) AXIOM B'. *Conditions and notations as above, if* $W^s(\Omega_i) \cap W^k(\Omega_j) \neq \emptyset$, *then there exist periodic orbits (or fixed points) γ in*

Ω_i, σ in Ω_j such that $W^s(\gamma)$ and $W^k(\sigma)$ have a point of transversal intersection.

The following seems to be a theorem although I haven't written out the details.[49]

(5.5) If ϕ_t: $M \to M$ satisfies Axioms A' and B' then ϕ_t is Ω-stable. One also obtains the openness, filtration, and partial ordering as in §I.6.

The approximation theorems are quite parallel to those referred to in §I.6 with the same references in fact.

(5.6) THEOREM [55] AND [114]. The property of flows that the fixed points x_i and closed orbits γ_j are all hyperbolic is generic. Furthermore generically, the stable and unstable manifolds of the x_i, γ_j intersect each other only transversally.

(5.7) THEOREM (PUGH [91]). In the Banach space of C' vector fields (or flows), there is a Baire set with the property that the fixed points and closed orbits are dense in Ω.

If ϕ_t: $M \to M$, ψ_t: $V \to V$ there is defined naturally the product flow $\phi_t \times \psi_t$: $M \times V \to M \times V$. Note that the product of two (or more) flows containing closed orbits of positive period will contain an invariant torus which will make this product not Ω-stable. For gradient flows (nondegenerate) the situation is different and simpler; the product is in this case Ω-stable.

Note that one obtains the example showing that structurally stable flows are not dense, by simply suspending the example for diffeomorphisms.

All the material in §I.5 about homoclinic points and symbolic flows can be suspended to obtain similar results on flows. As mentioned there, I first ran into this phenomena in that form, i.e., in trying to understand Van der Pol's equation (with forcing term). See also [107] for these questions discussed in the flow framework.

PART III. MORE ON FLOWS

III.1. **Flows with conditions imposed.** In this section, we discuss some of the problems encountered in attempting to carry over Parts I and II to flows which satisfy certain constraints, e.g., of the type occurring in classical mechanics. Essentially nothing has been done in this direction, so we just mention some background material, related recent results, and some problems.

The main class of flows, beyond the unrestricted ones we have been discussing up to this point, are the Hamiltonian flows. Abstractly speaking, a Hamiltonian flow is defined on a symplectic manifold, and this proceeds as follows.

A *symplectic* structure on a manifold M is a 2-form θ defined on M such that $d\theta = 0$, and at each point of M, θ is nondegenerate; nondegeneracy of θ at $x \in M$ means that the map $Q: T_x(M) \to T_x^*(M)$ is an isomorphism from the tangent space at x to its dual where $Q(X)(Y) = \theta(X, Y)$, $X, Y \in T_x(M)$ (for a complete discussion of this material, see [1], [123]). We then say that M is a symplectic manifold. It follows that dim M is even. Thus on a symplectic manifold, there is a 1-1 correspondence between 1-forms and vector fields.

Now if $H: M \to R$ is a differentiable function (a "Hamiltonian" function), its derivative $DH(x) \in T_x^*(M)$ defines a 1-form, which via Q we may consider as a vector field, say X_H. The flow ϕ_t generated by X_H (at least locally) is called the *Hamiltonian flow* defined by H. It can be checked that ϕ_t leaves θ invariant. In fact, by reasons converse to the above, it is important to consider directly those flows (which we will again call Hamiltonian) ϕ_t, say, defined for all t, on a symplectic manifold preserving the symplectic form. Then the natural global problem for Hamiltonian flows becomes

(1.1) PROBLEM.[50] Given a symplectic manifold M, find a Baire set $\mathfrak{B}$ of all flows which preserve the symplectic form, so that if ϕ_t is in $\mathfrak{B}$, one can describe the global orbit structure of ϕ_t.

If M is compact, one may conveniently consider the Hamiltonian flows $\mathcal{H}$ as a subspace of all vector fields, $\chi(M)$ with the C^r topology.

Note that a Hamiltonian flow, ϕ_t, leaves a volume on M invariant, namely the form obtained by wedging the symplectic form θ with itself $n = \frac{1}{2}$ dim M times. Thus it follows that in case M is compact, that the set of nonwandering points, Ω is equal to all of M.

One has a similar problem, also directly motivated by classical mechanics, for a single diffeomorphism.

(1.2) PROBLEM. What is the orbit structure of some Baire set of diffeomorphisms f of a compact symplectic manifold which preserve the symplectic 2-form?

Of course in studying these problems, one is only permitted perturbations of f to f' which also keep θ invariant. The first (and still unsolved) problem that one encounters here is to understand a local problem, the orbit structure in the neighborhood of a fixed point x of f. The difficulty is that the symplectic condition on f means that for the derivative $D(x)$, hyperbolicity is not a generic property. For example, if dim $M = 2$, f preserves a volume and $Df(x): T_x(M)$

$\to T_x(M)$ has determinant 1. One may classify these linear transformations into the hyperbolic and elliptic types. The hyperbolic is the one already discussed with eigenvalues $\lambda > 1$ and $1/\lambda$. The elliptic case is a nontrivial rotation of the plane. In the elliptic case in general, there exist no coordinates in the neighborhood of x in which f becomes linear, and only recently in this local 2-dimensional problem has one even begun to understand what is going on.

Birkhoff, e.g. [18], had believed that volume preserving transformations of compact 2-manifolds were ergodic (as well as Hamiltonian transformations more generally) "in the general case" and based much of his work on this hypothesis. (Recall ergodic means there are no invariant sets of positive measure with measure less than that of M.) Through the work of Kolmogoroff, Arnold, Moser, [52], [10], [71], we know now that this is not the case. If x is an elliptic fixed point of $C^\infty f: M^2 \to M^2$, then generically, there is an invariant circle·in every neighborhood of x and thus f cannot be ergodic [70].

In the $2n$-dimensional analogous problem there is an invariant n-dimensional torus in any neighborhood of x and the diffeomorphism is not ergodic. However one still has not yet a topological description in the neighborhood of an elliptic fixed point of a Hamiltonian diffeomorphism and thus it seems especially difficult to know how to proceed as in the first parts of the survey. Furthermore, the recent work of Arnold and Moser on the Hamiltonian case is still fairly local; the global Hamiltonian picture seems remote. We remark, though, that the examples of geodesic flows on manifolds of negative curvature are Hamiltonian and in this case, (§II.3), the flow is ergodic and structurally stable (on each level surface of the Hamiltonian).

We make three last comments on the Hamiltonian problem. First an elliptic point of a Hamiltonian diffeomorphism, say in 2 dimensions, where the derivative is a *rational* rotation, is degenerate. This is one reason why one must work with Baire sets of Hamiltonian diffeomorphisms, not open dense sets. Similarly one cannot expect these diffeomorphisms to be Ω-stable, as in Part I. Secondly, we remark that Pugh has shown that his closing lemma applies to prove the periodic orbits are dense in the compact Hamiltonian case [93]. Lastly it should be said that in practice, or in engineering, the differential equations, because of friction, are no longer Hamiltonian and could be closer to those described in Parts I and II. In this connection see [85].

After the Hamiltonian problems, the next most interesting case to consider might well be volume preserving diffeomorphisms. These coincide in dimension two with the Hamiltonian ones.

Volume preserving diffeomorphisms have not been studied from our point of view (although, see [122]).[51] For dim $M > 2$, however, none of the Hamiltonian objections apply and in fact the hyperbolic linear volume preserving maps are dense and open among all volume preserving linear maps; very possibly in the higher dimensional case, volume preserving diffeomorphisms might be amenable to study by the methods of Part I. A first question could be to prove I.(6.7) for volume preserving diffeomorphisms.[52]

For every volume preserving diffeomorphism f of a compact manifold, $\Omega(f) = M$. Presumably, Pugh's method would show the periodic points are dense. Is f ergodic, a generic property in this context? Oxtoby and Ulam [78] prove such an ergodicity theorem for homeomorphisms.

One can also ask whether the program of Parts I and II could be carried out for ordinary differential equations of higher order, say second order to begin with; see [56], [123] for a coordinate free definition of 2nd order differential equations. This hasn't been investigated as far as I know. The same applies to diffeomorphisms or flows of infinite dimensional manifolds.

Holomorphic diffeomorphisms of a complex manifold are much more rigid, but I think that the orbit structure is not generally understood. G. Julia's prize memoir [50] is related to this subject. It concerns holomorphic *endomorphisms* of the Riemann sphere.[53]

III.2. **Some other work on flows.** Here we mainly remark on a couple of recent results on flows which are not so directly related to the preceding.

The question of existence for minimal sets poses interesting problems to the global analyst. A compact manifold (or even space) M is a *minimal set* for the flow $\phi_t: M \to M$ if there is no proper nonempty closed invariant subspace of M. Gottschalk [35] has given a survey of this subject. A main problem is: what M can be the minimal set for some flow? It is not known if the 3-sphere can be a minimal set.[54]

A number of new examples of minimal flows are constructed from Lie groups in [12a]. See also [29] for examples on $S^n \times S^1$.

An important recent result is that of A. Schwartz [102] which generalizes both the Poincaré-Bendixson theorem for plane regions and Denjoy's theory of C^2 flows on the torus. The Schwartz theorem says that for any C^2 flow on a 2-manifold, any (compact) minimal set is either a point, a closed orbit, or a 2-torus. Among other applications of Schwartz's methods, R. Sacksteder has shown that if G is a finitely generated, finitely presented, discrete group G acting C^2

freely on the circle, then the action is topologically conjugate to a group of rotations. See also [99]. Here acting freely means no $\phi_g: S^1 \rightarrow S^1$, $g \in G$, has a fixed point.

There has been recently also interesting work on the subject of distal actions which we do not go into. Here $\phi_t: M \rightarrow M$, compact M, is distal, if in some metric, for any x, $y \in M$, $x \neq y$, there is an ϵ such that $d(\phi_t(x), \phi_t(y)) > \epsilon$ all $t \in G$. See for example [28], [33], [67].

PART IV. OTHER LIE GROUPS

IV.1. **Action of an abelian Lie group.** We consider briefly here the question of an abelian Lie group G acting on a manifold when G is more complicated than Z or R.

Recall first that an action of a Lie group G on a manifold is a homomorphism $\phi: G \rightarrow \mathrm{Diff}(M)$ such that the induced map $\Phi: G \times M \rightarrow M$ defined by $\Phi(g, m) = \phi_g(m)$ is C^k. The *orbit* O_x through $x \in M$ of such an action is the image of the map $p_x: G \rightarrow M$ defined by $p_x(g) = \phi_g(x)$. The *isotropy group* H_x of the action at x is the set of elements $h \in G$ such that $\phi_h(x) = x$. Then H_x is a closed subgroup of G and G/H_x is a homogeneous space of G. Induced from p_x is a 1-1 immersion $q_x: G/H_x \rightarrow M$. Finally we remark that the x orbit O_x refers to p_x, q_x, G/H_x or $q_x(G/H_x)$ at various times. If there is danger of confusion we will try to be more explicit.

A *fixed point* of the action is an orbit consisting of a single point. Actions ϕ_g, ψ_g are *conjugate* if there is a homeomorphism $h: M \rightarrow M$ such that $\phi_g(hx) = h(\psi_g(x))$ for all $g \in G$, $x \in M$.

Returning to the abelian case, suppose G is isomorphic to $Z + Z$. One may choose generators f, $g \in \mathrm{Diff}(M)$ of G, so $fg = gf$, and thus one is equivalently studying a pair of commuting diffeomorphisms. More generally one may study two commuting differentiable maps and actually the most studied of such problems perhaps has been the existence of a common fixed point for two commuting maps of the unit interval I into itself. Very recently a counterexample has been found to this problem by P. Huneke [49] and independently W. Boyce [22]. They each construct continuous maps f, $g: I \rightarrow I$, with $fg = gf$ and such that there is no $x \in I$ with $f(x) = x = g(x)$. These maps are not C^1 and thus the differentiable version of this problem remains open. In this direction, A. Schwartz [103] has the strongest result: If f and g are C^1 maps, $I \rightarrow I$, there is a fixed point of one which is periodic for the other.

Going back to the case of two commuting diffeomorphisms g, $f: M \rightarrow M$, observe g is in the centralizer $Z(f)$ of f, i.e., $Z(f) = \{g \in \mathrm{Diff}(M) | gf = fg\}$. Thus a first question in such a study could well be

(1.1) PROBLEM.[55] What can be said about $Z(f)$ for $f \in \mathrm{Diff}(M)$? Under what conditions on f is dim $Z(f) < \infty$? Is $Z(f) = \{f^m | m \in Z\}$ a generic property? .

Work of N. Kopell suggests that this last question may have an affirmative answer. Since a significant class of Ω-stable diffeomorphisms (§I.2) are a finite union of contractions up to a finite power, it is important to know $Z(f)$ when f is a contraction.

(1.2) THEOREM (KOPELL [53]). *Suppose* $C^\infty f: W \to W$ *is a contraction. Thus at the unique fixed point x, derivative* $Df(x): T_x(W) \to T_x(W)$ *is a linear contraction. Then* $Z(f) = \{g \in \mathrm{Diff}(W) | gf = fg, g \in C^\infty\}$ *is a finite dimensional Lie group. If f is linear, with a further nondegeneracy condition on the eigenvalues, then g is linear. Finally for a dense open set of diffeomorphisms f satisfying* I.(2.2), $Z(f) = \{f^m | m \in Z\}$.

In the proof of the structural stability of an Anosov diffeomorphism (see the Appendix of Part I), one obtains at the same time that its centralizer is discrete, even in the group of homeomorphisms of M. Adler and Palais [5] have actually computed this centralizer for the toral diffeomorphisms. It would seem at least a reasonable conjecture that an open dense set of diffeomorphisms satisfying Axioms A and B (§I.6) have centralizer $Z(f) = \{f^m | m \in Z\}$.[56]

Kopell [53] has studied commuting diffeomorphisms of the circle in more detail. Here, at least among those with periodic points, she has found a dense set of actions of $Z+Z$ for which the orbit structure can be understood. She also gives an example of commuting diffeomorphisms f, g of S' with the following property: g is the identity on an open set and for C' approximations f', g' of f, g such that $f'g' = g'f'$, g' must also be the identity on some open set.

Further results on abelian actions are related to the question of degeneracy of some orbits when R^k acts on a given manifold. By taking generators, an action of R^k on M corresponds to a set of k tangent vector fields on M which commute, or equivalently their bracket is zero.[57] In this direction Lima [60] showed that if R^2 acts on a compact 2-manifold of nonzero Euler characteristic, there must be a fixed point or, equivalently, a common zero of the two generating vector fields. In a further paper [59] he showed that two commuting vector fields on S^3 are dependent at some point (see also Novikov-Arnold [76]). Extensions of this last theorem have been made to actions of R^k on certain M^{k+1} by Rosenberg, and Sacksteder [98], [100]. While on this subject it seems worthwhile to mention that closely related is the result of Novikov [77] who has shown that every foliation of dimension 2 on S^3 has a compact leaf. An account of the basic results in foliation theory is in Haefliger [41].

Recently Adler and McAndrew [4] have shown that the topological entropy of a Chebyshev polynomial is positive.

There has not been much work on actions of solvable or nilpotent Lie groups along the line of this section.

IV.2. **The semisimple case.** Here we make a few comments on the problem of studying the action of a semisimple group G on a (compact) manifold. Discussions with R. Palais have been helpful here.

There is a vast literature on the subject of a Lie group G acting on a manifold M when G is compact, acts transitively, or acts linearly. The reader can refer to [20], [66] for the case of compact G. We only remark that Palais [80] (see also [81]) has shown a strong form of structural stability for compact actions. Namely if an action ψ_g is close to ϕ_g, $\phi: G \to \mathrm{Diff}(M)$, G compact, these actions are conjugate by a diffeomorphism $h \in \mathrm{Diff}(M)$. Thus $\phi_g(hx) = h\psi_g(x)$, all $g \in G$. In this case we say ϕ is *rigid*.

One systematic treatment of the transitive case is [27]. Another aspect of this case is [126].

If G acting on M is semisimple, but neither compact, nor acting transitively, nor linearly, there seems to be essentially no literature, at least that I know of.[58] On the other hand, it would seem worthwhile to make enorts in this direction. These efforts could produce unifying theorems, shed light on the above three special cases, or be useful in geometry or physics. One possibility might be to extend some of the results of Parts I and II. We limit ourselves to a few remarks.

In the first place, the evidence is that the richness of actions of a noncompact semisimple Lie group will lie somewhere between the abelian case (extremely rich, e.g., $G = R$) and the compact group case (few actions, i.e., G acts rigidly as mentioned above). We will try to make this point clearer.

In the linear theory, or representation theory, the semisimple case is close to the case of compact groups in that representations (finite dimensional) are rigid. This contrasts to the abelian case where even one dimensional representations of R (up to equivalence) are parameterized by R.

This motivated the speculation that if $\phi: G \to \mathrm{Diff}(M)$ is an action of semisimple G with fixed point $x \in M$, the representation $g \to D\phi_g$: $T_x(M) \to T_x(M)$ determines the orbit structure of ϕ in a neighborhood of x. R. Hermann [44] showed this to be true formally, while Guillemin and Sternberg [37] show that this is actually true in the case of an analytic (real) action. On the other hand Guillemin and Sternberg [37] give a counterexample in the C^∞ case for $G = \mathrm{SL}(2, R)$. This situation, however, is still not yet well understood.

One might ask whether any action of a semisimple G on a compact manifold is rigid. This is false as the following simple example shows. Let $G=\mathrm{SL}(2, R)$ act on the unit tangent bundle T of a 2-manifold M^2 of genus greater than one by dividing out a uniform discrete subgroup Γ from G. These actions correspond to different complex structures on M^2 and thus one gets a continuous family of such actions. One sees this by considering M as the double coset space $\Gamma\backslash G/K$ where G/K is the complex upper half plane. See also for example [126].

This does not exclude the possibility of a number of cases of noncompact semisimple G acting rigidly on compact M. Here is one such case. $G=\mathrm{SL}(n+1, R)$ acts transitively on $P^n(R)$ using homogeneous coordinates and in fact $\mathrm{SL}(n+1, R)$ has no other homogeneous spaces of dimension less than $n+1$. Thus there is at most one action, transitive or otherwise, of $\mathrm{SL}(n+1, R)$ on connected M if the dimension of M is less than $N+1$, so of course this action is rigid and M must be $P^n(R)$ (or a point!).

This suggests that semisimple G acting on M of much lower dimension might be fairly amenable to study. The situation is akin to the work of Hsiang and Hsiang on compact G [48].

The work of Hermann [43] and others on the (equivariant) compactification of homogeneous spaces may be interpreted as studying the action of a semisimple G in the neighborhood of certain noncompact orbits.

I have just received a manuscript [47] of W. Y. Hsiang which is related to the material of this section.

IV.3. **Final miscellany.** We end by making some final remarks on the action of a Lie group G. The notion of induced representation which has proved useful in the linear theory has an analogue in the general case which we describe now. This construction generalizes the suspended action of §II.1. Suppose then H is a closed subgroup of a Lie group G and $\phi\colon H\to\mathrm{Diff}(M)$ is an action of H. Define an action $\psi\colon H\to\mathrm{Diff}(M\times G)$ by $\psi_h(m, g)=(\phi_h(m), gh)$ and let $\pi\colon M\times G\to E$ be the projection onto the orbit space. One obtains the following diagram where f is induced by π_G.

$$
\begin{array}{ccc}
M\times G & \xrightarrow{\;\pi_G\;} & G \\
\downarrow & & \downarrow \\
E & \xrightarrow[\;f\;]{} & G/H
\end{array}
$$

Here E is a manifold and $f\colon E\to G/H$ is a bundle over G/H, Cartan's construction in [23]. The action σ of G on $M \times G$ defined by $\sigma_{g'}(m, g)$ $= (m, g' g)$ induces an action $\tau = \tau(\phi)$ of G on E which we might call the induced action of ϕ. This action $\tau: G \to \text{Diff}(E)$ is fiber preserving with respect to f and commutes with translation on the base. τ restricted to H leaves $f^{-1}(eH)$ invariant where it is equivalent to the original action ϕ.

If $G = R$, $H = Z$, then τ is the construction of the suspension of a generator of Z as in §II.1. If M is compact so is E and if ϕ is linear, M a vector space, then E is a vector space bundle and the action of G on sections is Mackey's induced representation.

We saw in Parts I and II that the concept of nonwandering points played a central role. A most important task would be to generalize this idea to a more general group G and to formulate some of the conditions say of §§I.2, I.3, I.6 for the case of a general Lie group, or even abelian, or semisimple G.[59]

Palais in [79] considers a class of actions on noncompact groups which have many properties of compact transformation groups. These actions are quite restrictive in that the isotropy group is always compact and the manifold must be noncompact. These actions, however, resemble $G = Z$ acting on $M - \Omega$.

Suppose now that $\phi: G \to \text{Diff}(M)$ is an action with a fixed point $x \in M$. Then the map $\tilde{\phi}: G \to \text{Aut}(T_x(M))$ is a linear representation of G, where $\tilde{\phi}(g) = D\phi_g(x)$ is the linear automorphism of $T_x(M)$ defined by the derivative of ϕ_g at x.

(3.1) PROBLEM. To what extent (generically) does this representation determine the action of G in a neighborhood of x, say up to conjugacy.

This is a basic local question. In earlier sections we saw aspects of it, starting with the stable manifold theorems, §I.2, in Sternberg [121] and Guillemin-Sternberg [37]. In general, the question is very far from being answered. Very likely, the higher derivatives will play a basic role for some groups.

Suppose more generally that $O = O_x$ is the compact orbit of some $x \in M$ of the action $\phi: G \to \text{Diff}(M)$. Thus $G \to O_x$, $g \to \phi_g(x)$ is the orbit map with isotropy group $H_x \subset G$ acting on M leaving x fixed. The derivative $D\phi_g: T_0(M) \to T_0(M)$ defines a structure of a homogeneous vector space bundle (in the sense of [21]) on the restriction of the tangent bundle of M restricted to O. We may generalize (3.1) with

(3.2) PROBLEM. To what extent does the group of bundle automorphisms $D\phi_g: T_0(M) \to T_0(M)$ determine the action of G in a neighborhood of O?

Here of course (3.1) is the case O is a single point. For the notion of equivalence in (3.2), one might take orbit preserving homeomorphism. An earlier special case of (3.2) was discussed in §II.2, where $G = R$ and O was a closed orbit, i.e., the circle S^1.

For global actions of G on M, it is probably profitable to consider initially very restricted cases, for example, actions on 2-manifolds. In this case, the possible orbits are well known, see Mostow [72]. Another tractable case might be actions with only two orbits.

BIBLIOGRAPHY

1. R. Abraham and J. Marsden, *Foundations of mechanics*, Benjamin, New York, 1967.

2. R. Abraham and J. Robbin, *Transversal mappings and flows*, Benjamin, New York, 1967.

3. R. L. Adler, A. G. Konheim and M. H. McAndrew, *Topological entropy*, Trans. Amer. Math. Soc. 114 (1965), 309–319.

4. R. L. Adler and M. H. McAndrew, *The entropy of Chebyshev polynomials*, Trans Amer. Math. Soc. 121 (1966), 236–241.

5. R. L. Adler and R. Palais, *Homeomorphic-conjugacy of automorphisms on the torus*, Proc. Amer. Math. Soc. 16 (1965), 1222–1225.

6. A. Andronov and L. Pontryagin, *Systèmes grossiers*, Dokl. Akad. Nauk. SSSR 14 (1937), 247–251.

7. D .V. Anosov, *Roughness of geodesic flows on compact Riemannian manifolds of negative curvature*, Soviet Math. Dokl. 3 (1962), 1068–1070.

8. ———, *Ergodic properties of geodesic flows on closed Riemannian manifolds of negative curvature*, Soviet Math. Dokl. 4 (1963), 1153–1156.

9. ———, *Geodesic flows on compact Riemannian manifolds of negative curvature*, Trudy Mat. Inst. Steklov. 90 (1967) = Proc. Steklov Math. Inst. (to appear).

10. V. I. Arnold, *Proof of a theorem of A. N. Kolmogorov on the invariance of quasi-periodic motions under small perturbations of the Hamiltonian*, Russian Math. Surveys 18 (1963), 9–36.

11. V. I. Arnold and A. Avez, *Problèmes ergodiques de la mécanique classique*, Gauthier-Villars, Paris, 1966.

12. E. Artin and B. Mazur, *On periodic points*, Ann. of Math. (2) 81 (1965), 82–99.

12a. L. Auslander, L. Green, F. Hahn et al., *Flows on homogeneous spaces*, Princeton Univ. Press, Princeton, N. J., 1963.

13. A. Avez, *Ergodic theory of dynamical systems*, Vols. I, II, Univ. of Minneapolis, Minnesota, 1966, 1967.

14. P. Billingsley, *Ergodic theory and information*, Wiley, New York, 1965.

15. G. D. Birkhoff, *Dynamical systems*, Amer. Math. Soc. Colloq. Publ., vol. 9, Amer. Math Soc., Providence, R. I., 1927.

16. ———, *Surface transformations and their dynamical applications*, Acta Math. 43 (1920), 1–119.

17. ———, *On the periodic motions of dynamical systems*, Acta Math. So. (1927), 359–379. (See also Birkhoff's Collected Works.)

18. ———, *Nouvelles recherches sur les systèmes dynamique*, Mém. Pont. Acad. Sci. Novi Lyncaei 1 (1935), 85–216.

19. ———, *Dynamical systems with two degrees of freedom*, Trans. Amer. Math. Soc. 18 (1917), 199–300.

20. A. Borel, et al., *Seminar on transformation groups*, Princeton Univ. Press, Princeton, N. J., 1960.

21. R. Bott, *Homogeneous vector bundles*, Ann. of Math. (2) **66** (1957), 203–248.

22. W. Boyce, *Commuting functions with no common fixed point*, Abstract 67T-218, Notices Amer. Math. Soc. **14** (1967), 280.

23. H. Cartan, *Seminaire H. Cartan espaces fibrés et homotopie*, Mimeographed, Paris, 1949–50.

24. E. Coddington and N. Levinson, *Theory of ordinary differential equations*, McGraw-Hill, New York, 1955.

25. J. Dieudonné, *Foundations of modern analysis*, Academic Press, New York, 1960.

26. A. Dold, *Fixed point index and fixed point theorem for Euclidean neighborhood retracts*, Topology **4** (1965), 1–8.

27. E. Dynkin, *The maximal subgroups of the classical groups*, Amer. Math. Soc. Transl. (2) **6** (1957), 245–379.

28. R. Ellis, *Distal transformation groups*, Pacific J. Math. **8** (1958), 401–405.

29. ———, *The construction of minimal discrete flows*, Amer. J. Math. **87** (1965), 564–574.

30. L. E. Elsgolts, *An estimate for the number of singular points of a dynamical system defined on a manifold*, Amer. Math. Soc. Transl. (1) **5** (1962), 498–510 (originally 68 (1952)).

31. F. B. Fuller, *The existence of periodic points*, Ann. of Math. (2) **57** (1953), 229–230.

32. ———, *An index of fixed point type for periodic orbits*, Amer. J. Math. **89** (1967), 133–148.

33. H. Furstenberg, *The structure of distal flows*, Amer. J. Math. **85** (1963), 477–515.

34. I. M. Gelfand and S. V. Fomin, *Geodesic flows on manifolds of constant negative curvature*, Amer. Math. Soc. Transl. (2) **1** (1955), 49–66.

35. W. H. Gottschalk, *Minimal sets: An introduction to topological dynamics*, Bull. Amer. Math. Soc. **64** (1958), 336–351.

35a. W. H. Gottschalk and G. A. Hedlund, *Topological dynamics*, Amer. Math. Soc. Colloq. Publ., vol. 36, Amer. Math. Soc., Providence, R. I., 1955.

36. E. Grosswald, *Topics from the theory of numbers*, Macmillan, New York, 1966.

37. V. Guillemin and S. Sternberg, *On a conjecture of Palais and Smale*, Trans. Amer. Math. Soc. (to appear).

38. B. Halpern, *Fixed points for iterates*, (to appear).

39. P. Hartman, *Ordinary differential equations*, Wiley, New York, 1964.

40. J. Hadamard, *Les surfaces à courbures opposées e leur lignes géodèsiques*, J. Math. Pures Appl. **4** (1898), 27–73.

41. A. Haefliger, *Variétés feuilletées*, Ann. Scuola Norm. Sup. Pisa (3) **16** (1962), 367–397.

42. G. Hedlund, *The dynamics of geodesic flows*, Bull. Amer. Math. Soc. **45** (1939), 241–246.

43. R. Hermann, *Lie groups for physicists*, Benjamin, New York, 1966.

44. ———, *The formal linearization of a semisimple Lie algebra of vector fields about a singular point*, Trans. Amer. Math. Soc. (to appear).

45. E. Hopf, *Ergodic theory*, Springer, Berlin, 1937.

46. ———, *Statistik der geodätischen Linien in Mannigfaltigkeiten negativer Krümmung*, Ber. Verh. Sächs. Akad. Wiss. Leipzig **91** (1939), 261–304.

47. W. Y. Hsiang, *Remarks on differentiable actions of noncompact semi-simple Lie groups on Euclidean spaces*, Amer. J. Math. (to appear).

48. W. C. Hsiang, and W. Y. Hsiang, *Differentiable actions of compact connected classical groups*, Amer. J. Math. (to appear).

49. J. P. Huneke, *Two counterexamples to a conjecture on commuting continuous functions of the closed unit interval*, Abstract 67T-231, Notices Amer. Math. Soc. **14** (1967), 284.

50. G. Julia, *Mémoire sur l'iteration des fonctions rationnelles*, J. Math. Pures Appl. **4** (1918), 47–245.

51. S. Kobayashi and J. Eells, *Problems in differential geometry*, Proc. Japan-United States Seminar in Differential Geometry, Kyoto, Japan 1965, Nippon Hyoransha, Tokyo, 1966, pp. 167–177.

52. A. N. Kolmogoroff, *On conservation of conditionally periodic motions for a small charge in Hamilton's function*, Dokl. Akad. Nauk SSSR **98** (1954), 527–530. (Russian)

53. N. Kopell, Thesis, Univ. of California, Berkeley, 1967.

54. B. Kostant, *Lie algebra cohomology and the generalized Borel-Weil theorem*, Ann. of Math. (2) **74** (1961), 329–387.

55. I. Kupka, *Contribution à la théorie des champs génériques*, Contributions to Differential Equations **2** (1963), 457–484.

56. S. Lang, *Introduction to differentiable manifolds*, Wiley, New York, 1962.

57. S. Lefschetz, *Differential equations: Geometric theory*, Interscience, New York, 1957.

58. N. Levinson, *A second order differential equation with singular solutions*, Ann. of Math. (2) **50** (1949), 127–153.

59. E. Lima, *Commuting vector fields on S^3*, Ann. of Math. (2) **81** (1965), 70–81.

60. ———, *Common singularities of commuting vector fields on 2-manifolds*, Comment. Math. Helv. **39** (1964), 97–110.

61. A. Malcev, *On a class of homogeneous spaces*, Amer. Math. Soc. Transl. (1) **9** (1962), 276–307 (originally **39** (1951)).

62. L. Markus, *Structurally stable differential systems*, Ann. of Math. (2) **73** (1961), 1–19.

63. Y. Matsushima, *On the discrete subgroups and homogeneous spaces of nilpotent Lie groups*, Nagoya Math. J. **2** (1951), 95–110.

64. F. Mautner, *Geodesic flows on symmetric Riemann spaces*, Ann. of Math. (2) **65** (1957), 416–431.

65. K. Meyer, *Energy functions for Morse-Smale systems*, Amer. J. Math. (to appear).

66. D. Montgomery and L. Zippen, *Topological transformation groups*, Interscience, New York, 1955.

67. C. Moore, *Distal affine transformation groups*, Amer. J. Math (to appear).

68. M. Morse, *The calculus of variations in the large*, Amer. Math. Soc. Colloq. Publ., vol. 18, Amer. Math. Soc., Providence, R. I., 1934.

69. ———, *A one to one representation of geodesics on a surface of negative curvature*, Amer. J. Math. **43** (1921), 33–51.

70. J. Moser, *On invariant curves of area-preserving mappings of an annulus*, Nachr. Akad. Wiss. Göttingen Math.-Phys. K1. II **1962**, 1–20.

71. ———, *Perturbation theory for almost periodic solutions for undamped nonlinear differential equations*, Internat. Sympos. Nonlinear Differential Equations and Nonlinear Mechanics, Academic Press, New York, 1963, pp. 71–79.

72. G. D. Mostow, *The extensibility of local Lie groups of transformations and groups on surfaces*, Ann. of Math. (2) **52** (1950), 606–636.

73. J. Nash, *Real algebraic manifolds*, Ann. of Math. (2) **56** (1952), 405–421.

74. V. V. Nemytskii, *Some modern problems in the qualitative theory of ordinary differential equations*, Russian Math. Surveys **20** (1965), 1–35.

75. K. Nomizu, *On the cohomology of compact homogeneous spaces of nilpotent Lie groups*, Ann. of Math. (2) **59** (1954), 531–538.

76. S. P. Novikov, The topology summer Institute, Seattle 1963, Russian Math. Surveys **20** (1965), 145–167.

77. S. P. Novikov, *Smooth foliations on three-dimensional manifolds*, Russian Math. Surveys **19** (1964) No. 6, 79–81.

78. J. C. Oxtoby and S. M. Ulam, *Measure-preserving homeomorphisms and metrical transitivity*, Ann. of Math. (2) **42** (1941), 874–920.

79. R. S. Palais, *On the existence of slices for actions of non-compact Lie groups*, Ann. of Math. (2) **73** (1961), 295–323.

80. ———, *Equivalence of nearby differentiable actions of a compact group*, Bull. Amer. Math. Soc. **67** (1961), 362–364.

81. R. S. Palais and T. Stewart, *Deformations of compact differentiable transformation groups*, Amer. J. Math. **82** (1960), 935–937.

82. J. Palis, Thesis, Univ. of California, Berkeley, 1967.

83. M. Peixoto, *On structural stability*, Ann. of Math. (2) **69** (1959), 199–222.

84. ———, *Structural stability on two-dimensional manifolds*, Topology **1** (1962), 101–120.

85. ———, Symposium on Differential Equations and Dynamical Systems (Puerto Rico) Academic Press, New York, 1967.

86. ———, *On an approximation theorem of Kupka and Smale*, J. Differential Equations **3** (1967), 214–227.

87. M. Peixoto and C. Pugh, *Structurally stable systems on open manifolds are never dense*, (to appear).

88. O. Perron, *Die stabilitatsfrage bei differentialgleichungen*, Math. Z. **32** (1930), 703–728.

89. H. Poincaré, *Sur les courbes dèfinies par des èquations différentielles*, J. Math. 4° serie 1885.

90. ———, *Les méthodes nouvelles de la mécanique céleste*, Vols. I, II, III, Paris, 1892, 1893, 1899; reprint, Dover, New York, 1957.

91. C. Pugh, *The closing lemma*, Amer. J. Math. (to appear)

92. ———, *An improved closing lemma and a general density theorem*, Amer. J. Math. (to appear).

93. ———, *The closing lemma for Hamiltonian systems*, Proc. Internat. Congress Math. at Moscow, 1966.

94. G. Reeb, *Sur certaines propriétés topologiques des trajectoires des systemes dynamiques*, Acad. Roy. Belg. Cl. Sci. Mem. Coll. 8° **27** (1952), no. 9.

95. L. Reiziņš, *On systems of differential equations satisfying Smale's conditions*, Latvijas PSR Zinātņu Akad. Vēstis Fiz. Tehn. Zinātņu Ser. **1964**, 57–61. (Russian)

96. F. Riesz, and B. Nagy, *Functional analysis*, Ungar, New York, 1955.

97. H. Rosenberg, *A generalization of Morse-Smale inequalities*, Bull. Amer. Math. Soc. **70** (1964), 422–427.

98. ———, *Actions of R^n on manifolds*, Comment. Math. Helv. **41** (1966–67), 170–178.

99. R. Sacksteder, *Foliations and pseudo-groups*, Amer. J. Math. **87** (1965), 79–102.

100. ———, *Degeneracy of orbits of actions of R^n on a manifold*, Comment. Math. Helv. **41** (1966–67), 1–9.

101. O. F. G. Shilling, *Arithmetical algebraic geometry*, Harper and Row, New York, 1956.

102. A. J. Schwartz, *A generalization of a Poincaré-Bendixon theorem to closed two-dimensional manifolds*, Amer. J. Math. **85** (1963), 453–458.

103. ————, *Common periodic points of commuting functions*, Michigan Math. J. **12** (1965), 353–355.

104. A. Schwartz, *Vector fields on a solid torus* (to appear).

105. H. Seifert, *Closed integral curves in 3-space and isotopic two-dimensional deformations*, Proc. Amer. Math. Soc. **1** (1950), 287–302.

106. A. Selberg, *Harmonic-analysis and discontinuous groups in weakly symmetric Riemann spaces with applications to Dirichlet series*, J. Indian Math. Soc. **20** (1956), 47–87.

107. L. P. Shil'nikov, *The existence of an enumerable set of periodic motions in the neighborhood of a homoclinic curve*, Dokl. Akad. Nauk SSSR **172** (1967), 298–301 = Soviet Math. Dokl. **8** (1967), 102–106.

108. M. Shub, Thesis, Univ. of California, Berkeley, 1967.

109. S. Smale, *Morse inequalities for a dynamical system*, Bull. Amer. Math. Soc. **66** (1960), 43–49.

110. ————, *On gradient dynamical systems*, Ann. of Math. (2) **74** (1961), 199–206.

111. ————, *On dynamical systems*, Bol. Soc. Mat. Mexicana (2) **5** (1960), 195–198.

112. ————, *Dynamical systems and the topological conjugacy problem for diffeomorphisms*, Proc. Internat. Congress Math. (Stockholm, 1962), Inst. Mittag-Leffler, Djursholm, 1963, pp. 490–496.

113. ————, *A structurally stable differentiable homeomorphism with an infinite number of periodic points*, Proc. Internat. Sympos. Non-linear Vibrations, vol. II, 1961, Izdat. Akad. Nauk Ukrain SSR, Kiev, 1963.

114. ————, *Stable manifolds for differential equations and diffeomorphisms*, Ann. Scuola Norm. Sup. Pisa (3) **17** (1963), 97–116.

115. ————, *Diffeomorphisms with many periodic points*, Differential and Combinatorial Topology, Princeton Univ. Press, Princeton, N. J., 1965, pp. 63–80.

116. ————, *Structurally stable systems are not dense*, Amer. J. Math. **88** (1966), 491–496.

117. ————, *Dynamical systems on n-dimensional manifolds*, Symposium on differential equations and dynamical systems (Puerto Rico) Academic Press, New York, 1967.

118. J. Sotomayor, *Generic one-parameter families of vector fields on two-dimensional manifolds* (to appear).

119. P. R. Stein and S. M. Ulam, *Non-linear transformations studies on electronic computers*, Rozprawy Mat. **39** (1964).

120. S. Sternberg, *Local contractions and a theorem of Poincaré*, Amer. J. Math. **79** (1957), 787–789.

121. ————, *On the structure of local homeomorphisms of Euclidean n-space. II*, Amer. J. Math. **80** (1958), 623–631.

122. ————, *The structure of local homeomorphisms. III*, Amer. J. Math. **81** (1959), 578–604.

123. ————, *Lectures on differential geometry*, Prentice-Hall, Englewood Cliffs, N. J., 1964.

124. R. Thom, *Sur une partition en cellules associés à une function sur une variété*, C. R. Acad. Sci. Paris **228** (1949), 973–975.

125. A. Weil, *Adeles and algebraic groups*, Notes, Institute for Advanced Study, Princeton, N. J., 1961.

126. ————, *On discrete subgroups of Lie groups. II*, Ann. of Math. (2) **75** (1962), 578–602.

127. R. Williams, *One dimensional non-wandering sets* (to appear).

NOTES

[1] These footnotes and references are in no way meant to be complete; I will give some of the more important results which bear directly on *Differentiable dynamical systems* (DDS) and have appeared since.

Let me first note some of the more comprehensive general references. Especially pertinent is the book by Shub, 1978, which gives a very good development of some of the main results of differentiable dynamical systems. There are also good accounts by Nitecki, 1971 and Palis–Melo, 1978. More elementary texts which give an ordinary differential equations background are Arnold, 1973 and Hirsch–Smale, 1974.

Then there are three survey articles by Shub; Shub, 1974, 1976, and in Peixoto, 1973, which give a brief account of recent research. Bowen's monograph, Bowen, 1978, is also of this nature. See also Chapter 7 in Abraham–Marsden, 1978, Markus, 1971, and "Fifty Problems by Palis–Pugh in Manning 1975.

Finally, one should mention various proceedings of conferences which contain research and survey articles on dynamical systems. These include Chern–Smale, 1970, or simply C–S, Manning, 1975, Markley et al., 1978, Nitecki–Robinson, 1980, and Peixoto, 1973.

[2] Such a goal is unachievable. Subsequent examples of Abraham and Smale, Newhouse, Simon, Shub, Williams, etc. have shown some of the difficulties. See footnote 23 and Chapter 3 of this volume for some perspectives of this problem.

[3] In place of the nonwandering set it is often convenient to use the chain-recurrent set, defined and developed by Conley and Easton. See, e.g., Conley, 1976, 1978 or Franke–Selgrade, 1977.

[4] The Morse inequalities have been generalized to cases with an infinite number of periodic orbits by Franks, 1975, 1978a and Zeeman in Manning, 1975. See also footnote 4.

[5] Peixoto (in Peixoto, 1973) explicitly has dealt with labeled diagrams (in the context of flows). There is much other work that deals implicitly or explicitly with Problem (2.4). See footnote 21.

[6] Problem (2.5) (as well as its counterpart for flows) has been solved affirmatively. These diffeomorphisms are structurally stable, Palis and Smale in C–S. See also Palis, 1969. Subsequently this result was generalized in the structural stability theorem of Robbin, 1971. See footnote 29.

[7] Shub in Peixoto, 1973 showed that M–S diffeomorphisms induce unipotent automorphisms of the homology groups. Then some deep work of the converse was done by Shub–Sullivan, 1975 and Franks–Shub, 1980.

Besides this is the work of Franks–Narisimhan, 1978, Narasimhan, 1979, Rocha, 1978, and Maller, 1979.

[8] The best weakening of the transversality condition (3) is to the no-cycle condition due to Rosenberg cited in DDS. See also "The Ω-stability theorem" (Smale) in C–S.

[9] The problem of whether $\Omega = M$ for an Anosov diffeomorphism is still open, but the corresponding problem for flows has been solved negatively by the important counter-example of Franks–Williams, 1979.

[10] The problem of finding all Anosov diffeomorphisms is also unsolved. But Franks in C–S solved the codimension-one case when $\Omega = M$, and Newhouse, 1970, removed the condition $\Omega = M$. Then Manning, 1974 solved the problem in case the manifold is a torus of some dimension. See also Auslander and Scheuneman in C–S.

Farrel and Jones, 1978b have found a series of new examples which are topologically conjugate to the hyperbolic infra-nil manifold automorphisms in Franks, ibid. These examples are on exotic tori and not diffeomorphic to infra-nil manifolds.

[11] As I remarked in "The Ω-stability theorem," C–S, this example is incorrect.

[12] Related to this section are the pseudo-Anosov diffeomorphisms of surfaces of Thurston, cf. Fathi et al, 1979. Also one can see Anosov–Sinai 1967 and Pugh–Shub 1972 for newer proofs of Anosov's ergodicity theorem.

[13] This is false by work of Simon, 1972. But see footnote 3 of I.6 for some positive results on the rationality of zeta functions. Also Mañé remarks (unpublished) that the Artin–Mazur theorem is a corollary of Bezout's theorem and the persistence of normally hyperbolic submanifolds.

[14] For a more recent exposition of symbolic dynamics, homoclinic points, and horseshoes see Moser, 1973. For a survey of the same, see Devaney in Gurel–Rössler, 1979. In both one finds reference to the work of Alekseev in celestial mechanics where ideas like this are applied.

Three examples of much recent literature in various fields where horseshoes are uncovered are Holmes–Marsden, 1980, Kopell–Howard, 1979, and Rössler, 1977.

Katok, 1979, in an important paper, shows how frequent this kind of dynamics is. E.g., he shows that in many homotopy classes of surface transformations, every diffeomorphism must have transversal homoclinic points.

[15] For more on the historical background, see Chapter 9 of this book.

It is interesting to see how Levi, 1978 has gone back to the equation of Cartwright and Littlewood which inspired Levinson's paper. It was in trying to geometrize Levinson's work that I found the horseshoe. Levi in Nitecki–Robinson, 1980 writes that the Cartwright–Littlewood equation, which arose in electronics, was a major incentive in the development of the theory of dynamical systems. Levi then applied this theory back to their equation.

[16] Good general references for recent perspectives on this section are Bowen, 1978 and Shub, 1978.

[17] This has been answered. Dankar, 1978 has shown that Axiom A(a) and Axiom A(b) are independent. Earlier Newhouse and Palis in Peixoto, 1973 had shown that Axiom A(b) is a consequence Axiom A(a) in case M has dimension two.

Malta, 1978 shows that if the Birkhoff center is hyperbolic with no cycles, then Axiom A and the no-cycle condition are satisfied.

[18] Indeed, this has been proved to be the case: the zeta function is rational for an Axiom A diffeomorphism. In C–S. Bowen and Bowen–Lanford prove this if the basic set Ω_i has dimension 0. Next, Williams, 1968 (see also Williams in C–S) showed that the zeta function is rational in case Ω_i is an attractor. Then Guckenheimer, 1970 proved the result for Axiom A no-cycle diffeomorphisms. Finally, Manning, 1971 proved generally that Axiom A implies rationality of the zeta function. Shub, 1978 has an exposition of a proof using Markov partitions.

[19] Axiom B is obsolete. It is replaced by either the no-cycle condition (see "The Ω-stability theorem" (Smale) in C–S or the strong transversality condition (see Palis–Smale C–S or Robbin, 1971), both of which are more natural than Axiom B was.

[20] See the paper "The Ω-stability theorem" (Smale) in C–S. Also important in this development is the 4-person paper Hirsch et al, 1970, but see especially Bowen, 1975, and his shadowing lemma.

[21] See footnotes 4, 5, and 7 whose references bear on this problem. There is also my result in Peixoto, 1973 that every diffemorphism is isotopic to an Axiom A no-cycle one. Then there are these relevant works: Batterson, 1979, Franks, 1977, 1979b, Maller, 1979, and unpublished work of David Fried and Joan Birman–Bob Williams.

[22] An extension of Pugh's results to C^2 approximations remains an important open problem.

[23] This is shown to be impossible by Abraham–Smale in C–S. This is reinforced by further examples of Newhouse, 1974, Simon, 1972, and Williams in C–S. There is the further example of the Lorenz attractor: See Guckenheimer–Williams, 1979, and Williams, 1979. See also footnote 2.

[24] See footnotes 30, 33, 34.

[25] See footnote 20.

[26] The basic reference to the generalized stable manifold theorem is Hirsch–Pugh–Shub, 1977. Important extensions of stable manifold theory are due to Pesin, 1979 for an area-preserving diffeomorphism and then to Ruelle, 1980 in the general case. I understand that Mañé has a book (in Portugese) published by IMPA in Rio de Janeiro on the theory of Pesin. Also in (7.3) one should replace "hyperbolic" by "Axiom A."

[27] For more on canonical coordinates, see Bowen in C–S. Also in (7.6)(a) X should be an orbit, not a point.

[28] For more on filtrations, see e.g. Nitecki–Shub: 1976, Shub in Peixoto, 1973, and Shub–Smale, 1972.

[29] The Ω-stability theorem asserts that an Axiom A diffeomorphism which has no cycles must be Ω-stable. The converse to the Ω-stability theorem seems plausible and remains an outstanding open problem. This amounts to showing that Ω-stability implies Axiom A, since Palis in C–S has shown that Axiom A and Ω-stability implies the no-cycle condition.

If a diffeomorphism satisfies Axiom A and the strong transversality condition (which is stronger than the no-cycle condition), then Robbin, 1971 has proved that this diffeomorphism is structurally stable. The converse to this again remains an outstanding open problem and is a consequence of: structural stability implies Axiom A. The last fact uses work of Robinson in Peixoto, 1973. Actually, in these conjectures in view of Newhouse, 1972, one only needs to confirm Axiom A(a).

We remark here that Robinson, 1976 has extended Robbin's theorem from C^2 hypotheses to C^1 and Robinson, 1974 makes the extension to vector fields. Pugh–Shub, 1970 prove the Ω-stability theorem for flows.

There have been some results on the converse problems, sufficient conditions for Axiom A. In particular, partial results are in Franks, 1972, 1974, Guckenheimer, 1972b: Lopez, 1979, Mañé in Manning, 1975, Mañé, 1978 and, Pliss, 1972.

Finally, very recently, Liao, 1979 and Mañé, 1980 seem to have shown that structurally stable diffeomorphisms in dimension 2 satisfy Axiom A.

[30] A Markov partition is a basic tool for understanding basic sets. See Sinai, 1968 for the case of an Anosov diffeomorphism and Bowen, 1975 for the general Axiom A case.

[31] The zeta function is indeed rational. See footnote 18.

[32] A basic set is not necessarily the product of a Cantor set and an interval even locally. This is the content of Guckenheimer's example in C–S.

[33] The analysis of Axiom A attractors has been developed much in the work of Williams, 1967, 1973. See also the recent papers of Farrel–Jones, 1979a, 1979b.

It is interesting to see that a 1-dimensional Axiom A attractor can occur already in a diffeomorphism of the 2-sphere, according to Plykin, 1974.

[34] These questions are answered by Bowen in C–S. The zero-dimensional basic sets comprise precisely the subshifts of finite type. See Williams, 1973 on the classification of these.

[35] The DE example is exposed in detail in Shub, 1977. See also my paper in Ratiu–Bernard, 1977. This example was used by Ruelle–Takens, 1971.

[36] The DA example was developed by Williams in C–S.

[37] The answer is yes. Topological entropy is positive for a (nontrivial) Axiom A basic set. This is proved by Bowen in C–S.

The entropy conjecture of Shub, 1974 has become the focus of much nice work in dynamical systems. The idea is that complexity on the homology level implies dynamical complexity. More precisely the entropy conjecture asserts that for any diffeomorphism and any eigenvalue λ of the induced map on homology, the topological entropy is greater than or equal to $\log |\lambda|$.

Partial results are due to Bowen, 1974, Fried–Shub, 1979, Manning, 1975, Misiurewicz–Przytycki, 1977, Ruelle–Sullivan, 1975, and Shub–Williams, 1975. See Bowen, 1978 for details.

Manning, 1979 discusses entropy for flows. Katok, 1979 obtains strong conclusions from positive topological entropy.

[38] Shub's thesis, published as Shub, 1969 introduces the examples of "infranil" expanding maps. The problem as to whether every expanding map is topologically conjugate to one of these examples has been recently solved by Gromov, 1980 in a fundamental paper. However, besides Shub's paper there had been partial results due to Franks and Hirsch, both in C–S.

Farrel–Jones, 1978a have found expanding maps on manifolds not diffeomorphic (but homeomorphic) to infranil manifolds.

[39] See Hirsch, 1971 on this topic.

[40] The work of Fatou and Julia on the iteration of complex analytic (i.e., rational) endomorphisms of the Riemann sphere has been revived by Brolin, 1965, Guckenheimer in C–S, Jewell, 1979, and Jakobson, M., 1969, among others.

Complex analytic ordinary differential equations have recently been studied by Guckenheimer, 1972a, Camacho–Kuiper–Palis, 1978, Iljasenko, 1969, 1977, and others.

[41] There are certain other areas which relate to DDS, but so much has been done recently that I can hardly begin to give references.

(1) Maps and diffeomorphisms of the interval and the circle: Here one must mention the important papers of Herman, 1977, 1979. See Bowen, 1978 and Guckenheimer, 1979 for some mention of the literature on other aspects of these problems.

(2) Bifurcation theory: I just mention these general references: Guckenheimer, 1979, Gurel–Rössler, 1979, Marsden–McCracken, 1976, and Palis, 1977, 1980.

(3) Numerical work: I believe there is much important work on this topic related to DDS.

(4) Ergodic theory: See Bowen, 1975, Ruelle, 1976, 1978, and Sinai, 1972, 1976. Also see Chapter 8 of this book.

[42] Fried, 1976 gives a nice condition on a flow for existence of a cross-section.

[43] I subsequently noticed that a conjecture of Birkhoff, 1950 (p. 710, problem 1), is answered by the construction of the suspension ([**114**] of DDS). Birkhoff asked whether any discrete flow might be imbedded in a continuous flow (as a section).

[44] Asimov, 1975a, 1975b has made a study of nonsingular flows satisfying (2.2). In particular, he showed that they exist on every manifold of dimension greater than three with nonzero Euler characteristic. Morgan, 1979 showed this is false in dimension 3.

Franks, 1978b, 1979a has also studied the inerterplay between the dynamics and the topology of these flows. He sharpened the Morse inequalities.

See also footnotes 4, 7 and 21.

[45] Peixoto's proof is invalid for a non-orientable 2-manifold, and that problem remains open. See Gutierrez, 1978.

The structural stability of flows satisfying (2.2) has been extended to n-dimensions. See footnotes 6 and 30.

S. SMALE

Franks–Williams, 1979 have shown that Ω is not necessarily M for an Anosov flow.

Tomter in C–S has analyzed a class of Anosov flows with $\Omega = M$ which are Lie group induced. I understand Handel and Thurston have found another example.

Results on Anosov flows have been obtained by Armendariz, 1979, Byers, 1972, Plante, 1972, Plante–Thurston, 1972, and Verjovsky, 1974.

[46] Seifert's problem has been settled, at least for C^1 flows, by Schweitzer, 1974. Schweitzer gives an example of a C^1 flow on a 3-sphere with no closed orbits. It is unknown if a C^2 example exists. See also footnote 54.

[47] See Bowen, 1978 for a survey of progress on this question. In particular, Gallovatti, 1976 has an example of a C^r, $r < \infty$, Axiom A, basic set, where $Z(s)$ has no meromorphic extension to C. On the other hand Ruelle, 1976 gives such an extension for Anosov flows with analyticity hypotheses.

[48] Another zeta function which counts closed orbits of certain flows is given by Fried, 1978. See also footnote 21 and, especially, Franks, 1979b.

[49] See Pugh–Shub, 1970 and a proof and see also footnote 30.

[50] There have been many results on this due to Robinson, Takens, and others. See Abraham–Marsden, 1978, Chapter 8, for a general reference. See also our paper, Chapter 8 of this volume.

[51] There has also been much done on measure-preserving diffeomorphisms and flows. See, for example, footnote number 5 (4) in I.10, and Anosov–Katok, 1970, where an ergodic volume-preserving diffeomorphism is constructed on essentially every manifold.

[52] See Robinson, 1979 for results on this problem.

[53] See footnote 40.

[54] The analogue for the discrete case has been solved. Katok and Fathi–Herman, 1977 have constructed a C^∞ minimal diffeomorphism of the 3-sphere. The problem as stated is still not solved.

[55] Partial results on this problem have been obtained by Andersen, 1976 and Palis, 1976. See the latter for a survey and see also Sad, 1979. The last sentence of Theorem (1.2) should be deleted.

[56] This problem is still open, but see Footnote 55.

[57] For R^k actions on manifolds, see the two articles by Camacho in Peixoto, 1973. Subsequent work on this was done by Kuiper, 1978, Camacho–Lins Neto, 1979, and G. Palis, 1978 and Camacho, 1978.

[58] See Schneider, 1974 for actions of $SL(2, R)$ and subsequent work by Uchida, 1980, for actions of $SL(n, C)$.

[59] See Pugh–Shub, 1975 as well as Hirsch, 1979 and Stowe, 1980.

REFERENCES FOR NOTES

Abraham, R. and Marsten, J., *Foundations of mechanics*, 2nd edition, Benjamin, Reading, Mass., 1978.

Andersen, B., *Diffeomorphisms with discrete centralizer*, Topology **15** (1976), 143–147.

Anosov, D. and Katok A., *New examples in smooth ergodic theory. Ergodic diffeomorphisms*. Trans. Moscow Math. Soc. **23** (1970), 1–35; Amer. Math. Soc., Providence, R.I., 1972.

Anosov, D. and Sinai, Y., *Some smooth ergodic systems*, Russ. Math. Surveys **22** (1967), 103–167.

Armendariz, P., Codimension one Anosov flows on manifolds with solvable fundamental group, (1979), preprint.

Arnold, V., *Ordinary differential equations*, MIT Press, Cambridge, Mass., 1973.

———, *Mathematical methods of classical mechanics*, Springer-Verlag, N.Y., 1978.

Asimov, D., *Round handles and non-singular Morse–Smale flows*, Annals of Math. **102** (1975a), 41–54.

———, *Homotopy of non-singular vector fields to structurally stable ones*, Annals of Math. **102** (1975b), 55–65.

Batterson, S., *Constructing Smale diffeomorphisms on compact surfaces*, (1979), preprint.

Birkoff, G., Some unsolved problems of theoretical dynamics, in Vol. II of *Collected mathematical papers of G. D. Birkoff*, Amer. Math. Soc., Providence, R.I., 1950.

Bowen, R., *Entropy versus homology for certain diffeomorphisms*, Topology **13** (1974), 61–67.

———, *Equilibrium states and the ergodic theory of Anosov diffeomorphisms*, Springer-Verlag, New York, 1975.

———, *On axiom A diffeomorphisms*, Amer. Math. Soc., Prov. R.I., 1978.

Brolin, H., *Invariant sets under iteration of rational functions*, Arkiv für Matematik **6** (1965).

Byers, W., *Some properties of Anosov flows*, Can. J. Math. **24** (1972), 1114–1121.

Camacho: C., *Structural stability theorems for integrable differential forms on 3-manifolds*, Topology **17** No. 2 (1978), 143–155.

Camacho, C., Kuiper, N. and Palis, J., *The topology of holomorphic flows with singularity*, Publ. Math. I.H.E.S. **48** (1978), 5–38.

Camacho, C., and Lins Neto, A., *The topology of integrable differential forms near a singularity*, (1979), preprint.

Chern, S. S. and Smale, S., *Global analysis*, Amer. Math. Soc. Providence, R.I., 1970 (referred to as C–S).

Conley, C., Some aspects of the qualitative theory of differential equations in *Dynamical systems*, Cesari, Hale, and LaSalle, ed., Acad. Press, N.Y., 1976.

———, Isolated invariant sets and the Morse index, Amer. Math. Soc., Providence, R.I., 1978.

Dankar, A., *On Smale's axiom A dynamical systems*, Ann. of Math. **107** (1978), 517–553.

Farrel, T., and Jones, L., *Examples of expanding endomorphisms on exotic tori*, Inventiones Math. **45** (1978a), 175–179.

———, *Anosov diffeomorphisms constructed from π_1 Diff(S^n)*, Topology **17** (1978b), 273–282.

————, *New attractors in hyperbolic dynamics*, Journ. Diff. Geom. (1979a), to appear.

————, *Expanding immersions on branched manifolds*, (1979b), preprint.

Fathi, A., et al, *Travaux de Thurston sur les surfaces*, Astérisque **66-67** (1979).

Fathi, A., and Herman, M., *Existence de difféomorphismes minimaux*, Astérisque **49** (1977), 37–59.

Franke, J., and Selgrade, J., *Hyperbolicity and chain recurrence*, Jour. Diff. Equations **26** (1977), 27–36.

Franks, J., *Differentiably Ω-stable diffeomorphisms*, Topology **11** (1972), 107–113.

————, *Time dependent stable diffeomorphisms*, Inventiones Math. **24** (1974), 163–172.

————, *Morse inequalities for zeta functions*, Ann. of Math. **102** (1975), 55–65.

————, *Constructing structurally stable diffeomorphisms*, Annals of Math. **105** (1977), 343–359.

————, *A reduced zeta function for diffeomorphisms*, Amer. Jour. Math. **100** (1978a), 217–243.

————, *The periodic structure of non-singular Morse–Smale flows*, Comm. Math. Helv. **53** (1978b), 279–294.

————, *Morse–Smale flows and homotopy theory*, Topology **18** (1979a), 199–215.

————, *Knots, links and symbolic dynamics*, (1979b), preprint.

Franks, J. and Narasimhan, C., *The periodic behavior of Morse–Smale diffeomorphisms*, Inventiones Math. **48** (1978), 279–292.

Franks, J. and Shub, M., *The existence of Morse–Smale diffeomorphisms* (1980), preprint.

Franks, J., and Williams: R. F., *Anomalous Anosov flows*, (1979), preprint.

Fried, David, Cross-sections to flows, Ph.D. Thesis, U. C. Berkeley, 1976.

————, *Flow equivalence, hyperbolic systems and a new zeta function for flows*, (1978), preprint.

Fried, D. and Shub, M., *Entropy, linearity, and chain recurrence*, Publ. Math. I.H.E.S. **50**, (1979), 203–214.

Gallovatti, G., *Zeta functions and basic sets*, Atti. Accad. Naz. Lincei Rend. **61** (1976), 309–317.

Gromov, M., *Groups of polynomial growth and expanding maps*, (1980), preprint, I.H.E.S.

Guckenheimer, J., *Axiom A + no cycles ⇒ $\zeta_f(t)$ rational*, Bull. Amer. Math. Soc. **76** (1970), 592–594.

————, *Hartman's theorem for complex flows in the Poincaré domain*, Compositio Math. **24** (1972a), 75–82.

————, *Absolutely Ω-stable diffeomorphisms*, Topology **11** (1972b), 195–197.

————, *Bifurcations of Dynamical Systems*, (1979), to appear.

Guckenheimer, J. and Williams, R. F., *Structural stability of Lorenz attractors*, Publ. Math. I.H.E.S. (1979).

Gurel, O.. and Rössler, O., *Bifurcation theory and applications in scientific disciplines*, New York Acad. of Sci., N.Y., 1979.

Gutierrez, C., *Structural stability for flows on the torus with a cross-cap*, Trans. Amer. Math. Soc. **241** (1978), 311–320.

Herman, M., *Measure de Lebesgue et nombre de rotation*, Lecture Notes in Math. **597** (ed. Palis and doCarmo), (1977), 271–293, Springer, N.Y.

Herman, M., *Sur la conjugaison différentiable des diffeomorphismes du cercle a des rotations*, Publ. Math. I.H.E.S. **49** (1979), 5–233.

Hirsch, M., *Anosov maps, polycyclic groups and homology*, Topology **10** (1971), 177–183.

———, *Stability of stationary points of group actions*, Annals of Math. **109** (1979), 537–544.

Hirsch, M., Palis, J., Pugh, C., and Shub, M., *Neighborhoods of hyperbolic sets*, Inventiones Math. **9** (1970), 121–134.

Hirsch, M., Pugh, C., and Shub, M., *Invariant manifolds*, Springer-Verlag, N.Y., 1977.

Hirsch, M., and Smale, S., *Differential equations, dynamical systems, and linear algebra*, Academic Press, N.Y., 1974.

Holmes, P., and Marsden, J., *A partial differential equation with infinitely many periodic orbits: Chaotic oscillations of a forced beam.* (1980), preprint.

Il'jasenko, J., *An example of equations $dw/dz = P_n(z, w)/Q_n(z, w)$ having a countable number of limit cycles and arbitrarily large Petrovski–Landis genus*, Mat. Sbornik **9** (1969), 365–378.

Il'yasenko, Y., *Remarks on the topology of singular points of analytic differential equations in the complex domain and Ladis' theorem*, Funct. Anal. and its Appl. **11**, No. 2 (1977), 105–113.

Jakobson, M. V., *On the problem of classification of polynomial endomorphisms of the plane*, Math. USSR Sbornik **9** (1969) #2.

Jewell, A., Rational maps of the Riemann sphere, Ph.D. Thesis, Univ. of Cal., Berkeley, 1979.

Katok, A., *Lyapunov exponents, entropy and periodic orbits for diffeomorphisms*, (1979), preprint.

Kopell, N., and Howard, L., *Target patterns and horseshoes from a perturbed central force problem: some temporally periodic solutions to reaction diffusion equations*, (1979), preprint.

Kuiper, N., La topologie des singularités hyperboliques des actions de R^2, (1978), preprint.

Levi, M., Qualitative analysis of the periodically forces relaxation oscillations, Ph.D. Thesis, N.Y.U., 1978.

Liao, S. T., *A basic property of a certain class of differential systems with some applications*, (1979), preprint.

Lopes, Artur, *Structural stability and hyperbolic attractors*. Trans. of Am. Math. Soc. **252** (1979), 205–219.

Maller, M., *Fitted diffeomorphisms of non-simply connected manifolds*, (1979), preprint.

Malta, I., *Hyperbolic Birkhoff centers*, (1978), preprint.

Mañe, R., untitled, (1980), preprint.

Mañe, R., *Contributions to the stability conjecture*, Topology **17**, (1978), 383–396.

Manning, A., *Axiom A diffeomorphisms have rational zeta functions*, Bull. Lond. Math. Soc. **3** (1971), 215–220.

————, *There are no new Anosov diffeomorphisms in tori*, Amer. Jour. Math. **96** (1974), 422–429.

————, ed., *Dynamical systems—Warwick 1974*, Springer-Verlag, N.Y., 1975.

————, *Toral automorphisms, topological entropy and the fundamental group*, Astérisque **50** (1977), 273–281.

————, *Topological entropy for geodesic flows*, Annals of Math. **110** (1979), 567–574.

Markley, N., Martin, J., and Perrizo, W., ed., *The structure of attractors in dynamical system*, Springer-Verlag, N.Y., 1978.

Markus, L., *Lectures in dynamical systems*, Amer. Math. Soc., Providence, R.I., 1971.

Marsden, J. and McCracken, M., *The Hopf bifurcation and its applications*, Springer-Verlag, N.Y., 1976.

Misiurewicz, M., and Przytycki, *Topological entropy and degree of smooth mappings*, Bull. Acad. Polon. Sci., Sér. Sci. Mat. Astronom. Phys. **25** (1977), 573–578.

Morgan, J. *Non-singular Morse–Smale flows on 3-dimensional manifolds*, Topology **18** (1979), 41–53.

Moser, J., *Stable and Random motion in dynamical systems*, Princeton Univ. Press, Princeton, N.J., 1973.

Narasimhan, C., *The periodic behavior of Morse–Smale diffeomorphisms on compact surfaces*, Trans. Amer. Math. Soc. **248** (1979), 145–169.

Newhouse, S., *On codimension-one Anosov diffeomorphisms*, Amer. Jour. Math. **92** (1970), 761–770.

————, *Hyperbolic limit sets*, Trans. Amer. Math. Soc. **167** (1972), 125–150.

————, *Diffeomorphisms with infinitely many sinks*, Topology **12** (1974), 9–18.

Nitecki, Z., *Differentiable dynamics*, M.I.T. Press, Cambridge, 1971.

Nitecki, Z., and Robinson, C., *Global theory of dynamical systems, Northwestern 1979*, Springer-Verlag, N.Y., 1980.

Nitecki, Z. and Shub, M., *Filtrations, decompositions, and explosions*, Amer. Jour. Math. **97**, (1976), 1029–1047.

Palis, G., *Linearly induced vector fields and R^2-actions on spheres*, Jour. Diff. Geom. **13** (1978), 163–190.

Palis, J., *On Morse–Smale dynamical systems*, Topology **8** (1969), 385–405.

————, *Rigidity of the centralizers of diffeomorphisms and structural stability of suspended foliations*, Springer-Verlag, N.Y., 114–121, 1976.

————, Some developments on stability and bifurcation of dynamical systems, in *Geometry and Topology* (ed. Palis and do Carmo), Springer–Verlag, N.Y., 1977.

Palis, J., *Moduli of stability and bifurcation theory*, Proc. Int. Cong. of Math., Helsinki, (1980), 835–839.

Palis, J. and Melo, W., *Introducão dos sistemas dinâmicos*, I.M.P.A., Rio, 1978.

Peixoto, M., Ed., *Dynamical systems*, Academic Press, N.Y., 1973.

Pesin, Ja, *Characteristic Lyapunov exponents and smooth ergodic theory*, Russian Math Surveys **32** (1979), 55–114.

Plante, J., *Anosov flows*, Amer. Jour. Math. **94** (1972), 729–754.

Plante, J. and Thurston, W., *Anosov flows and the fundamental group*, Topology **11** (1972), 147–150.

Pliss, V. A., *A hypothesis due to Smale*, Diff. Equat. **8** (1972), 203–214.

Plykin, R., *Sources and Sinks for A-diffeomorphisms*, Mat Sb. **23** (1974), 233–253.

Pugh, C. and Shub, M., *The Ω stability theorem for flows*, Inventiones Math. **11** (1970), 150–158.

————, *Ergodicity of Anosov Actions*, Inventiones Math., **15** (1972), 1–23.

————, *Axiom A actions*, Inventiones Math. **29** (1975), 7–38.

Ratiu, T. and Bernard, P., ed., *Turbulence seminar, Berkeley 1976/77*, Springer-Verlag, N.Y., 1977.

Robbin, J., *A structural stability theorem*, Annals of Math. **94** (1971), 447–493.

Robinson, C., *Generic properties of conservative systems*, Amer. Jour. Math. **XCII** (1970), 562–603.

————, *Structural stability of vector fields*, Annals of Math. **99** (1974), 154–175.

————, *Structural stability of C^1 diffeomorphisms*, J. Diff. Equations **22** (1976), 28–73.

Rocha, L. F. da, Caracterizacão das classes de isotopiã Morse–Smale em superficies, (1978), preprint.

Rössler, O., *Horseshoe-map chaos in the Lorenz equation*, Physics Letters **60A** (1977).

Ruelle, D., *A measure associated with axiom A attractors*, Amer. Jour. Math. **98** (1976), 619–654.

————, *Zeta functions for expanding maps and Anosov flows*, Inventiones Math. **34** (1976), 231–242.

————, *Thermodynamic formalism* Addison-Wesley, Reading, Mass., 1978.

————, Ergodic theory of dynamical systems, Publ. Math. I.H.E.S. **50** (1979), (to appear).

Ruelle, D., and Sullivan, D., *Currents, flows and diffeomorphisms*, Topology **14** (1975), 319–327.

Ruelle, D. and Takens, F., *On the nature of turbulence*, Comm. Math. Phys. **20** (1971), 167–192; **23** (1971), 343–344.

Sad, P., *Centralizers of vector fields*, Topology **18** (1979), 97–104.

Schneider, C., SL(2, *R*) *actions on surfaces*, Amer. Jour. Math. **96** (1974), 511–528.

Schweitzer, P., *Counterexamples to the Seifert conjecture and opening closed leaves of foliations*, Ann. of Math. **100** (1974), 386–400.

Shub, M., *Endomorphisms of compact differentiable manifolds*, Amer. Jour. Math. **91** (1969), 175–199.

——, *Dynamical systems, filtrations, and entropy*, Bull. Amer. Math. Soc. **80** (1974), 27–41.

——, The lefschetz fixed-point formula; smoothness and stability in *Dynamical systems* (ed. Cesari, Hole, and LaSalle) Academic Press, N.Y., 1976.

——, Stability in dynamical systems in *Systèmes dynamiques et modèles économiques*, (ed. Fuchs, G. and Munier, B.) CNRS, Paris; 1977.

——, *Stabilité globale des systemes dynamiques*, Asterisque **56** (1978).

Shub, M., and Smale, S., *Beyond hyperbolicity*: Annals of Math. **96** (1972), 587.

Shub, M., and Sullivan, D., *Homology theory and dynamical systems*, Topology **14** (1975), 109–132.

Shub, M., and Williams, R. F., *Entropy and Stability*, Topology **14** (1975), 329–338.

Simon, C., *Instability in* Diff(T^3) *and the non-genericity of rational zeta functions*, Trans. Amer. Math. Soc. **174** (1972), 217–242.

Sinai, Ja, *Construction of Markov partitions*, Functional Anal. Appl. **2** (1968), 245–253.

——, *Gibbsian measures in ergodic theory*, Russian Math Surveys **27** (1972), 21–69.

——, *Introduction to ergodic theory*, Princeton Univ. Press, Princeton, N.J., 1976.

Stowe, D., *The stationary set of a group action*, Proc. Amer. Math. Soc. (1980), to appear.

Uchida, F., *Real Analytic* SL(*n*, *C*) *actions*, Bull. of Yamagato Univ. **10** (1980), 1–14.

Verjovsky, A., *Codimension one Anosov flows*, Bol. Soc. Mat. Mexicana **19** (1974), 49–77.

Williams, R. F., *One dimensional non-wandering sets*, Topology **6** (1967), 473–487.

——, The zeta function of an attractor in *Conference on the topology of manifolds* (*Michigan State 1967*), Prindle Weber and Schmidt, Boston, 1968.

——, *Expanding attractors*, Publ. Math., I.H.E.S. **43** (1973a), 169–203.

——, *Classification of subshifts of finite tyre*, Ann. of Math. **98** (1973b), 120–153 and errata **99**, 380–381.

——, *The structure of Lorenz attractors*, Publ. Math., I.H.E.S. **50** (1979), 73–99.

WHAT IS GLOBAL ANALYSIS?

There has recently been a lot of activity in that branch of mathematics now referred to as "global analysis." For example, the subject of the 1968 Summer Institute of the American Mathematical Society was global analysis.

My definition of global analysis is simply the study of differential equations, both ordinary and partial, on manifolds and vector space bundles. Thus one might consider global analysis as differential equations from a global, or topological point of view.

Even the earliest studies of differential equations contained an element of global analysis; this element had become quite important for example in the work of Poincaré on ordinary differential equations. G. D. Birkhoff's development of dynamical systems and especially M. Morse's theory of geodesics are both excellent examples of global analysis. After the rapid recent progress in topology, the subject of our exposition has been moving especially fast. After mentioning a couple of references in partial differential equations, I shall devote the rest of my article to an account of a theorem in dynamical systems to illustrate the global analysis point of view.

Recently there have been nice results in the topology of linear elliptic differential operators, especially in the work of Atiyah, Singer, and Bott (see for example [2] and [4]).

One cannot expect to have a satisfactory framework for nonlinear partial differential equations with linear function spaces. Thus it is important that nonlinear partial differential equations are beginning to be attacked by a systematic use of infinite dimensional manifolds of maps. A good survey of this is Eells [3].

The work of Andronov, Pontryagin [1] and Peixoto [5] in dynamical systems (or ordinary differential equations), on one hand can be explained in relatively simple terms and on the other hand gives a real insight into this modern way of looking at differential equations. I shall try to give a brief account of their theory now.

Consider an ordinary differential equation (1st order, autonomous) defined on a domain D in the x, y-plane:

$$\frac{dx}{dt} = P(x, y) \qquad \frac{dy}{dt} = Q(x, y).$$

Stephen Smale received his Ph.D. at the University of Michigan in 1956. He has occupied various positions at the University of Chicago, the Institute for Advanced Study, Columbia University, and his present location, the University of California at Berkeley. For his outstanding research in differential topology and in global analysis, Professor Smale was awarded the Fields Medal of the International Mathematical Union in 1966 and the Veblen Prize of the American Mathematical Society in 1964. *Editor*

4

We shall assume that these functions P, Q defined on D are continuously differentiable (or of class C^1). Now the fundamental existence theorem of ordinary differential equations yields for each (x_0, y_0) in D and real t sufficiently small in absolute value, $|t| < \epsilon$, functions $f(x_0, y_0, t)$, $g(x_0, y_0, t)$ which satisfy the initial conditions $f(x_0, y_0, 0) = x_0$, $g(x_0, y_0, 0) = y_0$ and the differential equation

$$\left(\frac{df}{dt}\right)(x_0, y_0, t) = P(f(x_0, y_0, t), g(x_0, y_0, t))$$

$$\left(\frac{dg}{dt}\right)(x_0, y_0, t) = Q(f(x_0, y_0, t), g(x_0, y_0, t)).$$

Let us look at this phenomenon from a more geometric point of view and in fact get away from the particular choice of x, y-coordinates.

To each (x, y) in D associate the vector $(P(x, y), Q(x, y))$ of the x, y-plane with the initial point at (x, y). This gives us what is called a C^1 vector field on D. For each point p of D, we will call the associated vector for short $X(p)$. Then the existence theorem we just stated may be interpreted to yield a system of plane curves $\phi_t(p)$ with $\phi_0(p) = p$, and with the property that the tangent of the curve at a point q of D will be the vector $X(q)$ (see Figure 1).

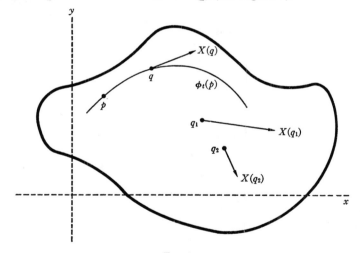

FIG. 1.

The right context for the study of this differential equation becomes clearer now. More generally than a domain of the Euclidean plane, consider a 2-dimensional smooth manifold M. Roughly speaking, one can think of this as a surface in 3-dimensional Euclidean space E^3 or better abstractly as a space on which differentiation makes sense and a neighborhood of each point is a domain in the plane. To each point p of M there is associated a 2-dimensional vector space

$T_p(M)$, the tangent space of M at p. If M is a surface in E^3 then $T_p(M)$ is the plane tangent to M at p.

A vector field X on M is an assignment, continuously differentiable, $p \rightarrow X(p)$ for p in M to $X(p)$, a "vector" in $T_p(M)$. The vector field on D defined previously from the differential equation given by the functions P, Q on D is now a vector field on the 2-manifold D in this sense.

To define the basic idea of this article, structural stability of a differential equation, we need to develop two things: one, the space of differential equations on M, $\chi(M)$, and two, an equivalence relation on $\chi(M)$, the phase portrait.

We have seen that the kind of differential equations on M we are studying (which are really pretty general except for the low dimension) correspond to vector fields on M. We call the set of all vector fields (C^1 as usual) on M, $\chi(M)$.

Now $\chi(M)$ has the structure of a vector space, using the fact that for each $p \in M$, the values of all vector fields lie in the same linear space $T_p(M)$. That is if X, Y belong to $\chi(M)$, $(X+Y)(p) = X(p) + Y(p)$. This space $\chi(M)$ will be basic in what follows.

The solution curves $\phi_t(p)$ of a vector field X on M, defined earlier, may be "pieced together " so that for each p, $\phi_t(p)$ will be defined for all $a < t < b$ where the interval (a, b) is maximal. If M is compact, for each p, this interval will be $(-\infty, \infty)$ so that we have a 1-parameter group ϕ_t of transformations on M. Thus for each real t, ϕ_t is a C^1 transformation of M, $\phi_t \colon M \rightarrow M$, with a C^1 inverse, ϕ_0 is the identity and $\phi_t(\phi_s) = \phi_{t+s}$. In short ϕ_t is a dynamical system.

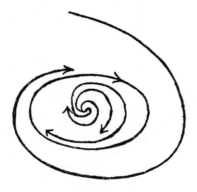

FIG. 2a.

To abstract the qualitative features of a differential equation on M, the concept of a phase portrait becomes important. Usually the phase portrait means the picture of the solution curves of the differential equation. For example, Figure 2a is the phase portrait of a differential equation in the plane.

To give a precise mathematical content to "phase portrait," we proceed as follows. Say X, Y in $\chi(M)$ are *topologically equivalent* when there is a homeomorphism $h: M \rightarrow M$ taking solution curves of X into those of Y. Thus the differential equation in Figure 2a is topologically equivalent to that described in Figure 2b.

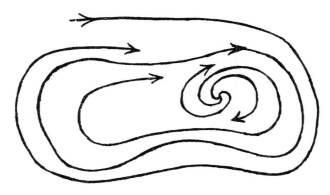

Fig. 2b.

Then two differential equations on M have the same phase portrait if they are topologically equivalent. A definition of *phase portrait* is thus a topological equivalence class of differential equations on M. A main goal of the qualitative study of ordinary differential equations is to obtain information on the phase portrait of differential equations.

To make progress in this direction, one soon sees the need to avoid "degenerate" cases. For example a differential equation that is zero on all of M, or even on some nonempty open set of M should be considered degenerate and excluded from most considerations. I think that engineers and physicists will agree with this statement.

To aid in discussing the question of degeneracy, a topology or metric on $\chi(M)$ is useful. To simplify matters in defining this metric, in the rest of our article, we will assume M compact. This excludes many or even most interesting examples, but on the other hand the main features are not lost.

Assuming M compact define a norm $\| \ \|$ on $\chi(M)$ as follows. Let U_1, $\cdots$, U_k be a covering of M, $\overline{U}_i \subset V_i$, with each V_i a plane domain. Then on each V_i, X in $\chi(M)$ is represented by $P_i(x, y)$, $Q_i(x, y)$ as at the beginning. Then $\|X\|$ is defined as the maximum of the following finite set of numbers:

$$\sup_{(x,y) \in U_i} | P(x, y) | \qquad i = 1, \cdots, k$$

$$\sup_{(x,y) \in U} | Q(x, y) | \qquad i = 1, \cdots, k$$

$$\sup_{(x,y)\in U_i} \left| \frac{\partial P}{\partial x}(x, y) \right| \qquad i = 1, \cdots, k$$

$$\sup_{(x,y)\in U_i} \left| \frac{\partial P}{\partial y}(x, y) \right| \qquad i = 1, \cdots, k$$

$$\sup_{(x,y)\in U_i} \left| \frac{\partial Q}{\partial x}(x, y) \right| \qquad i = 1, \cdots, k$$

$$\sup_{(x,y)\in U_i} \left| \frac{\partial Q}{\partial y}(x, y) \right| \qquad i = 1, \cdots, k.$$

This gives $\chi(M)$ the structure of a complete normed space or a Banach space. A metric on $\chi(M)$ is then defined by $d(X, Y) = \|X - Y\|$.

With this metric on $\chi(M)$ it is possible to say when differential equations are "close." In terms of local coordinate representations, two differential equations are close when the P and Q are uniformly close, with their first derivatives uniformly close as well.

With this background, we say that X in $\chi(M)$ is *structurally stable* when there is a neighborhood $N(X)$ in $\chi(M)$ with the property that every Y in $N(X)$ is topologically equivalent to X. Thus X is structurally stable when nearby differential equations have the same phase portrait. A little thought will indicate that this excludes degeneracy; a structurally stable X cannot be degenerate (in some senses at least). It is an important concept for the engineer who studies qualitative differential equations, since in engineering the differential equations one works with are only approximations of the real equations. The engineer wants the qualitative conclusion he makes to be valid for the actual differential equation which describes his world. In fact the original idea of structural stability was the joint work of an engineer, A. Andronov, and a mathematician, L. Pontryagin.

Thus it becomes important to know if most differential equations are structurally stable.

THEOREM. (M. Peixoto) *If M is a compact 2-dimensional manifold, then the structurally stable differential equations in $\chi(M)$ form an open and dense set.*

This theorem is an excellent theorem in global analysis. One sees in two ways how it is global. First the differential equation is defined over a whole manifold, and structural stability depends on its behavior everywhere. Second, the theorem makes a conclusion about the space of all differential equations on M.

The proof gives much information on the structure of differential equations on 2-manifolds.

We state the main lemma which indicates how this is so.

The nonwandering set $\Omega(X)$ of X is defined as the set of x in M such that for every neighborhood U of x and t_0, there is a $t > t_0$ with $\phi_t(U) \cap U \neq \varnothing$.

MAIN LEMMA. *If M is a compact 2-manifold and X is in $\chi(M)$, then X is structurally stable if and only if the following conditions are met:*

(a) *Each closed orbit and each singular point of X is "nondegenerate." This nondegeneracy is defined in terms of derivatives associated to the closed orbits and singular points.*

(b) *The separatrices of saddle points don't meet.*

(c) *$\Omega(X)$ consists of the finite union of closed orbits and singular points.*

[*Separatrices* are the trajectories which come to and leave from the saddle points.]

If α is a singular point or closed orbit, let $W^{\bullet}(\alpha)$ be the set of x in M with $\phi_t(x) \to \alpha$ as $t \to \infty$. Then if X is structurally stable, it provides for a decomposition of M as the finite union of $W^{\bullet}(\alpha)$ as α ranges over the closed orbits and singular points. This decomposition gives a good practical understanding of the differential equation X.

A survey of this subject with many references is [6].

This article is based on an address before the Mathematical Association of America, San Francisco, 26 January, 1968.

References

1. A. Andronov and L. Pontryagin, Systèmes grossiers, Dokl. Akad. Nauk. SSSR, 14 (1937) 247–251.

2. M. Atiyah and R. Bott, A Lefschetz fixed point formula for elliptic differential operators, Bull. Amer. Math. Soc., 72 (1966) 245–250.

3. J. Eells, A setting for global analysis, Bull. Amer. Math. Soc., 72 (1966) 751–807.

4. R. Palais *et al.*, Seminar on the Atiyah-Singer index theorem, Ann. of Math., Study No. 57, (1966).

5. M. Peixoto, Structural stability on 2-dimensional manifolds, Topology 1 (1962) 101–120.

6. S. Smale, Differentiable dynamical systems, Bull. Amer. Math. Soc., 73 (1967) 747–817.

Séminaire BOURBAKI   Février 1970
22ᵉ année, 1969/70, n° 374

STABILITY AND GENERICITY IN DYNAMICAL SYSTEMS

A general reference to this subject, with examples, written about the summer of 1967 is [7], (reported in a recent Bourbaki seminar by C. Godbillon). Here I will try to emphasize developments since. An important source of much of this more recent work should appear in the immediate future in [1].

For simplicity we restrict ourselves to a dynamical system defined by a diffeomorphism f of a compact manifold M into itself. This is the case of a discrete differentiable dynamical system with time represented by the number of times f is iterated or the n in f^n. Most of the results discussed here are valid also in the case of a dynamical system defined by a 1st-order ordinary differential equation.

The space of all dynamical systems on M will be denoted by $\mathrm{Dyn}(M)$, topologized by putting the C^r uniform topology on the corresponding diffeomorphism f, $1 \leqslant r \leqslant \infty$. The study of the dynamical system is the study of the orbits $O(x) = \{ f^n(x) \mid n \in Z \}$ of f especially from the global point of view. Thus a natural equivalence relation is topological conjugacy, i.e., f, $g \in \mathrm{Dyn}(M)$ are *topologically conjugate* if there is a homeomorphism $h : M \to M$ such that $fh = hg$. Clearly such an h sends the orbits of f onto the orbits of g.

More than 10 years ago I posed the problem of finding a dense open set U (or at least a Baire set) of $\mathrm{Dyn}(M)$ such that the elements of U could somehow be described qualitatively by discrete numerical and algebraic invariants. Since this problem has been often quoted since, I would like to take this opportunity to revise the problem in the light of what has been learned in these 10 years.

The problem posed in this way is too simple, too rough, and too centralized. I believe now that the main problems of dynamical systems can't be unified so elegantly. The above problem, however, can be split apart so that it makes good sense and in my opinion gives some perspective to the subject. This goes as follows.

One should search for a sequence of subsets U_i of $\mathrm{Dyn}(M)$, $U_1 \subset U_2 \subset \cdots \subset U_k \subset \mathrm{Dyn}(M)$, k not too large, U_i open (or at least, say, a Baire subset of an open set) with U_k dense in $\mathrm{Dyn}(M)$. One main feature of the U_i is that as i increases, U_i includes a substantially bigger class of dynamical systems, but as i decreases one has a deeper understanding and the elements of U_i have greater regularity (or stability properties). So U_1

should consist of the simplest best-behaved nontrivial class of dynamical systems, and U_k cannot be expected to have very strong stability properties at all.

To give a better idea of what I am saying, I will give a schema of the U_i now which to some extent illustrates our state of knowledge of dynamical systems. (There will always be some arbitrariness in the exact choice of the U_i.) We will first state briefly the choice of the U_i's and the rest of the talk will give some justification of our choice, defining the necessary terms as we proceed. In each of the following U_i there is a large class of examples not in the preceding U_{i-1}. The reader may consult the literature cited for many of these.

$U_1 = \{ f \in \text{Dyn}(M) \mid f \text{ satisfies Axiom A}, \ \Omega(f) \text{ is finite, and } f \text{ satisfies}$ the transversality condition$\}$;

$U_2 = \{ f \in \text{Dyn}(M) \mid f \text{ satisfies Axiom A and the strong transversality}$ condition$\}$;

$U_3 = \{ f \in \text{Dyn}(M) \mid f \text{ satisfies Axiom A and the no-cycle condition}\}$;

$U_4 = \{ f \in \text{Dyn}(M) \mid f \text{ satisfies the main known generic properties}\}$.

We have used the unifying language of Axiom A which can be stated as follows: $f \in \text{Dyn}(M)$ satisfies *Axiom* A if the nonwandering set $\Omega = \Omega(f)$ has a hyperbolic structure and the periodic points of f are dense in Ω. We recall that Ω is the closed invariant set of $x \in M$ such that for any neighborhood U of x there is some $n > 0$ with $f^n(U) \cap U \neq \emptyset$. A *hyperbolic structure* on Ω is a continuous splitting of the tangent bundle $T_\Omega(M)$ of M restricted to Ω, $T_\Omega(M) = E^u + E^s$, invariant under the derivative, Df, such that Df is contracting on E^s and $D(f^{-1})$ is contracting on E^u. Finally $Df : E^s \to E^s$ is said to be *contracting* if given a Riemannian metric on M, there is $c > 0$, μ, $0 < \mu < 1$ with $\| Df^m(x)(v) \| \leqslant c\mu^m \|v\|$ for all $v \in E^s$.

A *generic property* is a property that is true for some Baire set in $\text{Dyn}(M)$.

A basic notion for the study of dynamical systems is the notion of stable manifold. Given $f \in \text{Dyn}(M)$ and some fixed metric on M, we say that $x \sim_s y$ if $d(f^m(x), f^m(y)) \to 0$ as $m \to \infty$. This is an equivalence relation; the equivalence class of x is denoted by $W^s(x)$ and called the *stable manifold* of x. The following theorem is a consequence of the work of a number of mathematicians, see especially [1], [5].

THEOREM 1. *If* $f \in \text{Dyn}(M)$ *satisfies Axiom* A, *then for each* $x \in M$, $W^s(x)$ *is a smoothly injectively immersed open cell.*

It is an outstanding question as to whether the conclusion of Theorem 1 is a generic property.

The *unstable manifolds* of f are the stable manifolds of f^{-1} and are denoted by $W^u(x)$.

With this behind us consider the dynamical systems in U_1. That Ω is finite implies that Ω consists of the periodic points of f, and Axiom A amounts to saying that if $x \in \Omega$ has period m, then $Df^m(x) : T_x \to T_x$ has no eigenvalues of absolute value 1. The transversality condition of U_1 means that if $x \in M$, then $W^s(x)$ and $W^u(x)$ meet transversally at x. (Stated in this manner, this transversality condition coincides with the strong transversality condition of U_2.)

If $f \in U_1$, then f has indeed very strong stability properties. Say that $f \in \mathrm{Dyn}(M)$ is *structurally stable* if it possesses a neighborhood of diffeomorphisms, each topologically conjugate to f.

THEOREM 2 (PALIS–SMALE [1]). *If $f \in \mathrm{Dyn}(M)$ with $\Omega(f)$ finite, then f is structurally stable if and only if $f \in U_1$.*

It was known for some time via gradient dynamical systems that for any M, U_1 is not empty and more recently U_1 was shown to be open [6].

Structural stability is a very strong regularity property (now known to be not generic) and largely via the preceding theorem, U_1 can be considered to consist of very well understood dynamical systems of relatively simple character. On the other hand, the proof of this theorem is not altogether simple because of the fact that structural stability is such a strong (and subtle) property.

To define the remaining terms used in describing U_2, say that for $f \in \mathrm{Dyn}(M)$ satisfying Axiom A, f has the *strong transversality property* if for any $x \in M$, $W^s(x)$ and $W^u(x)$ meet transversally at x.

It has been conjectured that a necessary and sufficient condition for $f \in \mathrm{Dyn}(M)$ to be structurally stable is that $f \in U_2$. It is known that there exists $f \in U_2$, $f \notin U_1$, which is structurally stable. Among the rather complicated examples of $f \notin U_1$, some have the property that a neighborhood doesn't even intersect U_2 because of lacking the strong transversality property. Also for f satisfying Axiom A, it can be seen that the strong transversality property is necessary for f to be structurally stable. Via this route it was first found that structural stability was not a generic property. Proving that $U_2 =$ structurally stable dynamical systems would cement U_2 into our hierarchy.

We pass to U_3. To understand the no-cycle property, we recall the spectral decomposition theorem which states that if f satisfies Axiom A, then $\Omega(f)$ can be written canonically as the finite union of closed invariant disjoint subsets $\Omega_1, \ldots, \Omega_k$ on each of which f has a dense orbit.

Define $W^s(\Omega_i) = \bigcup_{x \in \Omega_i} W^s(x)$ and $W^u(\Omega_i) = \bigcup_{x \in \Omega_i} W^u(x)$. A *cycle* is a sequence of distinct $\Omega_1, \ldots, \Omega_k$, $k > 1$, with the property $W^s(\Omega_i) \cap W^u(\Omega_{i-1}) \neq \emptyset$, $i = 2, \ldots, k$, and $W^s(\Omega_1) \cap W^u(\Omega_k) \neq \emptyset$. Then f (supposed to satisfy Axiom A) has the *no-cycle property* if there are no cycles.

Dynamical systems in U_3 have the important regularity property known as Ω-stability which is defined as follows. First say that f, $g \in \mathrm{Dyn}(M)$ are Ω-*conjugate* or *conjugate on* Ω if there is a homeomorphism $h : \Omega(f) \to \Omega(g)$ such that $hf(x) = gh(x)$ for all $x \in \Omega(f)$. Then f is Ω-*stable* if it has a neighborhood N in $\mathrm{Dyn}(M)$ of diffeomorphisms which are Ω-conjugate to f. Clearly structurally stable implies Ω-stable.

THEOREM 3 (THE Ω-STABILITY THEOREM). *If $f \in U_3$, then f is Ω-stable.*

The converse of Theorem 3 is an open problem. More generally one can ask what stability properties of dynamical systems are valid outside of U_3. Can even some dynamical sytem not in U_3 be structurally stable? Another version of these questions is: Does structural stability imply Axiom A? or even does Ω-stability imply Axiom A? My feeling is that the questions of this paragraph are very hard and important to settle.

Some other regularity properties are true of $f \in U_3$. For example one can define the "zeta function" $\zeta(f) = \sum_{n=1}^{\infty} N_n t^n$, where N_n is the number of fixed points of f^n. It is an open question whether $\zeta(f)$ having a positive radius of convergence is a generic property. But on the other hand, the following theorem was very recently proved by J. Guckenheimer [4].

THEOREM 4. *If $f \in U_3$, then $\zeta(f)$ not only has a positive radius of convergence, but it is a rational function whose zeros and poles are algebraic numbers.*

R. Bowen [1], [2], [3] has studied dynamical systems satisfying Axiom A in the direction of ergodic theory and has obtained the following rather striking results.

THEOREM 5. *Let f satisfy Axiom A and Ω_i be one of the subsets given by the spectral decomposition theorem. Then there exists an invariant ergodic measure μ_f on Ω_i, positive on open sets, zero on points (unless Ω_i is finite) which is the unique invariant normalized Borel measure on Ω_i maximizing entropy. The (measure-theoretic) entropy coincides with the topological entropy and this entropy is equal to the log of the radius of convergence of the zeta function of f.*

Also Bowen gets good information on the distribution of periodic points in Ω_i and shows that $(f|_{\Omega_i}, \mu_f)$ is a K-automorphism in the "C-dense" case, a mild condition which is met, for example, in the case Ω_i is connected.

Let me emphasize again that indeed these last theorems cover situations which are very rich in examples and that I am not giving them here.

What happens outside of U_3? At the present time, there are a large number of examples outside of U_3 whose import is that one cannot obtain any dense open $U_k \subset \mathrm{Dyn}(M)$ with very strong regularity or stability properties. Some of them are as follows:

Abraham–Smale in [1] show that Ω-stability and Axiom A are not generic properties. Shub has an example of an open set in $\mathrm{Dyn}(M)$, where $\Omega = M$, where Axiom A and Ω-stability fail. Newhouse in [1] shows that if $r > 1$, with the C^r topology, even on S^2, Axiom A and Ω-stability are not generic properties. The earlier examples, with r arbitrary in the range $1 \leqslant r \leqslant \infty$, were on higher-dimensional manifolds. C. Simon has recently shown that the zeta function being rational is not a generic property.

Under perturbation, some features of these examples seem to be preserved. On the other hand the above examples and others need much study to get some understanding of the area beyond U_3.

As far as I know, there has been not much progress in the finding of new generic properties (or study of U_4) since [7] was written.

REFERENCES

[1] *Global Analysis*, Proceedings of the 1968 AMS institute on global analysis at Berkeley. Vol. 1 (of 3 vols) is devoted to dynamical systems.

[2] R. Bowen, Periodic points and measures for Axiom A diffeomorphisms, *Trans. AMS* **154** (1971), 377–397.

[3] R. Bowen, Markov partitions for Axiom A diffeomorphisms, *Amer. J. Math.* **92** (1970), 725–747.

[4] J. Guckenheimer, Axiom A + no cycles $\Rightarrow \zeta_f(t)$ rational, *Bull. AMS* **76** (1970), 592–594.

[5] M. Hirsch, J. Palis, C. Pugh, M. Shub, Neighborhoods of hyperbolic sets, *Inventiones Math.* **9** (1970), 121–134.

[6] J. Palis, On Morse–Smale dynamical systems, *Topology* **8** (1969), 385–405.

[7] S. Smale, Differentiable dynamical systems, *Bull. AMS* **73** (1967), 747–817.

PERSONAL PERSPECTIVES ON MATHEMATICS AND MECHANICS

1. These days especially, provocative questions confront a socially conscious scientist when he begins to contemplate where applications of his work might lead. As one whose main work has been in pure mathematics, and who is beginning to concern himself with areas of applied mathematics such as electrical circuit theory, I wonder to what extent I should explicitly direct my work toward socially positive goals. Many issues on the relations of the profession of a scientist to the social crises of this time deserve a format, for example, at various professional meetings and conferences, as well as in the departments of universities. There are strong pressures on the scientist, from the sources of funding research, scientific tradition, and a basic conservatism of maintaining the status quo, which act to prevent discussion of these issues. Although I am not going to pursue such a discussion here, I feel that mathematicians and scientists in general must face these questions in a much more serious way than they have done to date (myself included).

2. A fairly general procedure for the mathematical study of a physical system starts with explication of the space of *states* of that system. Now this space of states could reasonably be one of a number of mathematical objects. However, in my mind, a principal candidate for the state space should be a differentiable manifold; and in case the system has a finite number of degrees of freedom, then this will be a finite dimensional manifold. Usually associated with the physical system is the notion of how a state progresses in time. The corresponding mathematical object is a dynamical system or a first order ordinary differential equation on the manifold of states.

Too often in the physical sciences, the space of states is postulated to be a linear space, when the basic problem is essentially nonlinear; this confuses the mathematical development.

On the other hand, recent decades have seen big developments in parts of mathematics related to differentiable manifolds. Thus there should be no reason for reticence on the part of applied mathematicians to use these developments.

For the International Union of Pure and Applied Physics Conference on Statistical Mechanics, Chicago, March 1971.

We have tried to provide a little background for what follows. I will describe briefly aspects of three areas in which I have been working. These are mechanics, differentiable dynamical systems, and electrical circuits.

3. Recent years have seen several textbook accounts of the development of classical mechanics on manifolds. Two of these are Abraham and Marsden, Foundations of Mechanics [1], and Loomis and Sternberg, Advanced Calculus [2]. A number of journal articles also have appeared on aspects of this subject, including my "Topology and Mechanics" [3].

A basic motivation of this literature is to make the discipline of classical mechanics more elegant, more geometric (and so perhaps less analytic), more global. Furthermore, this development has done a lot to unify mechanics with large parts of geometry and part of topology. The globalization has removed the need for a linear space framework, such as one finds in the traditional classical mechanics text.

I will give a quick indication how this goes. Configuration space is a smooth manifold M. For the space of states one takes the tangent bundle T of M or the cotangent bundle T^* of M. The manifold T^* has a natural symplectic structure, i.e., a canonical nondegenerate 2-form Ω defined on it. This form defines an isomorphism between vector field and 1-forms. Thus if $H : T^* \to R$ is a function on T^* (a "Hamiltonian"), dH is a 1-form which corresponds to a unique vector field X_H via Ω. A vector field is the same thing as a first order ordinary differential equation; in fact, X_H is one way of representing Hamilton's equations. It is immediate then that $X_H(H) = 0$ or that differentiation of H along X_H is zero. This is conservation of energy. Symmetry and other conservation laws are related similarly, simply and conceptually. Furthermore, X_H preserves Ω and hence the volume element Ω^n, where $n = \dim M$. This is Liouville's theorem.

Suppose one starts with a kinetic and potential energy as well as M. How does that relate to this context? Potential energy is a smooth function $V : M \to R$, and kinetic energy can be thought of as a Riemannian metric on M with $K : T \to R$ given by norm squared of a tangent vector with respect to this Riemannian metric. Total energy $E : T \to R$ is the function $K + V o \Pi$, where $\Pi : T \to M$ is the projection. The Riemannian metric induces a bundle isomorphism (the Legendre transformation $T \to T^*$) and thus pulls Ω back to T to give a symplectic structure on T. Thus there is a vector field X_E naturally associated to E via this symplectic structure. The vector field X_E generalizes Newton's equations, and of course the energy E is conserved. If one has a Lie group $\mathfrak{G}$ of symmetries acting on M leaving E invariant, one has a natural induced map α of the Lie algebra $\mathcal{Y}$ of $\mathfrak{G}$ into vector fields on M. Let $\alpha_x : \mathcal{Y} \to T_x$ be the linear map obtained by evaluating the vector field $\alpha(X)$ at $X \in M$ where $X \in \mathcal{Y}$. One then has a map $J_1 : T^* \to \mathcal{Y}^*$ obtained by letting J_1 restricted to a fiber T_x^* be the dual linear map of α_x. The composition $J, T \to T^* \to \mathcal{Y}^*$ generalizes angular momentum. This map $J : T \to \mathcal{Y}^*$ is constant on orbits, and furthermore it can be shown that J is equivariant with respect to the adjoint action on G on $\mathcal{Y}^*$. This description of angular momentum helps clarify Jacobi's "elimination of the node" in celestial mechanics.

In the above if $V \equiv 0$, one has simply the case of Riemannian geometry, and the projections into M of the integral curves of X_E are geodesics. More generally the projections of these integral curves are the trajectories in configuration space.

We have given here the flavor of how this modernization of mechanics goes. The assertions above can usually be proved with very little effort, if one has accepted the basic elements of differential topology and geometry. The above-mentioned three references do cover and expand on this material.

For many modern mathematicians, the elegance of this treatment of mechanics would be justification for its development. Beyond this, some new insights have been obtained already by this globalization with, I believe, much more to come.

4. A lot of research has been done in recent years in the area of mathematics now called differentiable dynamical systems. Let me describe briefly, roughly, what it is about and after that give an assessment. For more substantial and precise information on this subject one can see my "survey," "Differentiable Dynamical Systems" [4], and its references, as well as the more recent Global Analysis [5].

The discipline of differentiable dynamical systems attempts to study systematically dynamical systems on a manifold. So one forms the topological space Dyn(M) (with a topology respecting differentiable properties) of all dynamical systems on a manifold M.

One has a reasonable notion of "almost all" in Dyn(M) meaning a Baire set (a countable intersection of open dense sets), and when convenient one restricts attention to such a set. A theorem is regarded as useful if, for example, it gives a property valid for all systems of some Baire set of Dyn(M).

Related is the idea of structural stability and certain variations. This kind of stability is a property of a dynamical system itself (not of a state or orbit) and asserts that nearby dynamical systems have the same structure. The "same structure" can be defined in several interesting ways, but the basic idea is that two dynamical systems have "the same structure" if they have the same gross behavior, or the same qualitative behavior. For example, the original defintion of "same structure" of two dynamical systems was that there was an orbit preserving continuous transformation between them. This yields the definition of structural stability proper. It is a recent theorem that every compact manifold admits structurally stable systems, and almost all gradient dynamical systems are structurally stable.

But while there exists a rich set of structurally stable systems, there are also important examples which are not structurally stable, and have good but weaker stability properties.

One series of theorems has been in the direction of characterizing various stability properties of dynamical systems in terms of more tangible properties of the system. One cannot expect this development per se to help in the study of mechanical systems as in section 3 above, because small perturbations will destroy the Hamiltonian behavior of a mechanical system: in particular, dissipation accomplishes this. On the other hand, engineering, and problems from the biological sciences, could profit from this approach through stability questions. What happens near an attractor or sink would be of special interest in applications. Most simple is the steady-state case (fixed point of the dynamical system), then the oscillation (or periodic behavior); but recent studies have shown how easy it is to have far more complicated attractors appear. In particular, even the structurally stable attractors in rather simply given examples contain a vast mixture of periodic, almost periodic, homoclinic, and other kinds of phenomena.

So differentiable dynamical systems have developed as a purely mathematical subject: the attempt to study all ordinary differential equations on a manifold. As such, a wealth of theorems and knowledge has been accumulated. On the other hand, the subject has not achieved any definitive state.

One of the main recent achievements has been to remove the fear of making a global study of differential equations on manifolds of dimension greater than 2. Many examples, and in fact large classes of systems, with good stability properties are now known and well understood on manifolds of all dimensions. On the other hand, we have concrete examples not understood as to their stability properties; in fact, there remain good questions to be found. The subject is in a state of activity and flux at this time. As I have said, up to this point the study of differentiable dynamical systems has developed as a purely mathematical discipline with only minimal direct relations to physics and engineering. I suspected there is a good potential for interaction. For example, the above-described work might be used to extend the horizons of applied mathematics by expanding the realm of possibilities for mathematical models.

5. I would like to give here the flavor of some recent work of mine on electric circuits—from "Mathematical Foundations of Electrical Circuits," preprints in existence. Our development proceeds in a natural manner, working with what is given by the circuit and making no arbitrary choices. We consider circuits—for example, as in C. Desoer and E. Kuh, Basic Circuit Theory [6]—with nonlinear elements, either resistors, inductors, or capacitors. Associated with a circuit is a linear graph, say connected and oriented, consisting of nodes or 0-cells and elements, branches, or 1-cells.

An unrestricted state is the set of currents and voltages in each branch and is thus a $2b$-tuple of real numbers, where b is the number of branches. The space $\mathfrak{S}$ of all states then is real Cartesian space of $2b$ dimensions. Kirchoff's and Ohm's laws will impose conditions so that a *physical* state must lie in a certain subset $\sum \subset \mathfrak{S}$ which we proceed to define.

First, Kirchoff's current and voltage laws define relations on the currents and voltages in the branches, and thus restrict states to lie in a lineaer subspace $\mathfrak{K}$ of $\mathfrak{S}$.

Let $i_R \times v_R : \mathfrak{S} \to R \times R'$ be the projection defined by taking the resistance currents and resistance voltages. Thus R consists of currents through the resistors and R' is the linear space of voltages across the resistors. We take Ohm's law to say that for each resistor ρ, the characteristic is a closed one-dimensional submanifold Λ_ρ of the current voltage plane of that resistor. Then these Λ_ρ define a closed submanifold (their product) Λ in $R \times R'$. Let $\Pi' : K \to R \times R'$ be the restriction of $i_R \times v_R$. Then consider.
Hypothesis: Π' is transversal to Λ.

Although this is a key hypothesis, rather than recalling its rather technical definition we note the main consequences that are of concern to us. Under this hypothesis, $\Pi'^{-1}(\Lambda) = \sum$ is a smooth manifold of $\mathfrak{S}$ and $\dim \sum =$ the number of inductors plus the number of capacitors. Furthermore, this hypothesis is generic in that it will be satisfied with almost all choices of resistor characteristics.

The space of *physical* states is naturally defined as $\sum$ (those satisfying both Kirchoff's and Ohm's laws). We see the above hypothesis as key, because in giving a manifold structure to $\sum$, it permits analysis of the space of states. In fact, via Faraday's laws, we can obtain an ordinary differential equation, perhaps singular, on $\sum$.

In brief, this goes as follows (but see the above reference for details). The inductances and capacitances define a symmetric form I over $\sum$; this form cannot, in general, be expected to be positive definite at each point of $\sum$. In fact, besides being indefinite, I may also be degenerate on part of $\sum$. The resistances define a closed 1-form w on $\sum$. Then the differential equations are the gradient of w with respect to I. These are going to be much more complicated than the usual gradients because of the general nature of I. These equations contain in increasing generality those of Van der Pol, Lienard, and Brayton-Moser. On the other hand, they contain examples which are of a different type corresponding to "relaxation oscillations" and thus contribute to achieving a certain unity in the differential equations of circuit theory.

6. There is a remaining aspect of this global analysis approach to applied mathematics that I would like to mention. The approach taken in this paper may indeed be a little hard to accept for the applied mathematician trained in traditional methods. However, the approach here tends to make applied mathematics and also ordinary differential equations accessible and attractive to the modern mathematician, one who has been brought up in the purist, Bourbaki style of education.

For several decades, the most important mathematical tendencies in the United States and Western Europe have been very separated from applications and there has even been a certain disdain for applied mathematics by many leading mathematicians. I believe that there have been some healthy sides to this separation. Many fields in pure mathematics have flourished in this time, and perhaps because of the division a certain clarity has been achieved.

On the other hand, at least since before the time of Newton and Leibniz, never have the main schools of mathematics, the main mathematicians, been so sharply separated from other disciplines. I believe it worthnhile to try to change this course of events. The modernizations I have described in mechanics, ordinary differential equations, and circuits will hopefully help to break this isolation by bringing modern mathematics more substantially into communication with other fields.

REFERENCES

1. Abraham and Marsden, Foundations of Mechanics (New York, 1967).
2. Loomis and Sternberg, Advanced Calculus (Reading, Mass., 1968), chap. 13.
3. Steve Smale, "Topology and Mechanics," I and II, Inventiones Math. 1970.
4. Steve Smale, "Differentiable Dynamical Systems," Bull. Amer. Math. Soc. 1967.
5. S. S. Chern and Steve Smale, Global Analysis, vol. 1 (Providence, R.I.: American Mathematical Society, 1970).
6. C. Desoer and E. Kuh, Basic Circuit Theory (New York, 1969).

DISCUSSION

J. M. Deutch: Have the techniques you discussed for structural stability and electrical networks been applied to systems of coupled, nonlinear chemical reactions?

M. E. Fisher: Would you make some comments on the "realizability" of a given circuit. By this I mean the question of some sort of existence theorem that shows that the model is "complete" and does not require the specification of further conditions, such as jump conditions, or regularization devices to avoid singularities, etc.

P. T. Landsberg: Returning to the first part of your talk, one could, of course, introduce Riemann geometry into classical mechanics and exhibit it in the guise of the mathematical framework of general relativity (see Frankman, for example). Is this the kind of generalization of classical mechanics which you have in mind? Indeed, is it general enough for your purposes?

R. Mazo: You have shown us how many familiar results of classical mechanics arise in your "modern" reformulation. I assume that the purpose of reformulating the theory is not only to express known results in new language but also to provide tools for obtaining new results. I wonder if you could give us an example of the sort of new results to which your formulation leads.

I. Prigogine: As this session is devoted to perspectives in mechanics, it seems not out of place to mention some recent work which shows that classical and quantum dynamics may be viewed in a new, rather unusual way as linear eigenvalue problems in eigenprobabilities.

Let us first consider the case of classical dynamics (for more detail see I. Prigogine, C. George, and J. Rae, Physica [1971, in press]). As is well known, a basic method for solving the integration problem in classical dynamics is through canonical transformation to a "cyclic" Hamiltonian. Then all momenta are constants. This is generally done through the solution of the Hamilton-Jacobi equation. Let us indicate an alternative way which leads to a linear eigenvalue problem. This is at first very surprising, but it is, in fact a direct consequence of the introduction of the "subdynamics formalism" in our recent papers (I. Prigogine, Cl. George, and F. Henin, Physica 45, 418 [1969], Proc. Nat. Acad. Sci. 65, 489 [1971]). Let us consider the Hamiltonian

$$H(J, \alpha) = H_0(J) + \lambda H_1(J, \alpha), \tag{1}$$

where

$$J = (J_1, \ldots, J_N), \alpha = (\alpha_1, \ldots, \alpha_N) \tag{2}$$

are action-angle variables (the generalization to other sets of canonical variables present no problem). To equation (1) corresponds the Liouville operator

$$L = L_0 + \lambda L_1 \tag{3}$$

defined in terms of the Poisson bracket $\{\ ,\ \}$

$$L_f = i\{H, f\}. \tag{4}$$

We introduce the projection operator P on the null space of L_0 together with its complement $\phi = 1 - \phi$ and the analogous projector π for the null space of L. Both P and π are self-adjoint operators. By definition each invariant I of the motion governed by H belongs to the null space of L so that

$$\pi I = I. \tag{5}$$

The operator π has remarkable properties implying a well-defined relation between the component (PI) in the null space of the unperturbed Hamiltonian H_0 and the complementary component ϕI:

$$\phi I = \mathcal{C}(PI) \tag{6}$$

where $\mathcal{C}$ is what we call a reaction operator (for its definition, see Prigogine et. al. 1969). Its exact form is of no importance here. It may be calculated explicitly in terms of J and derivatives $\partial^n / \partial J^n$. (The main point is that we now introduce a statistical ensemble characterized by the probability distribution $\tilde{\rho}$ and take the average of an invariant, say H, over the ensemble.)

Let us call $\tilde{J}$ the action variable corresponding to the complete Hamiltonian H such that

$$H = \tilde{H}(\tilde{J}). \tag{7}$$

We then have for the average value

$$\langle H \rangle = \frac{1}{(2\pi)^N} \alpha \tilde{J} \, d\tilde{\alpha} \tilde{H}(\tilde{J}) \tilde{\rho}(\tilde{J}, \alpha) = \int d\tilde{J} \tilde{H}(\tilde{J}) \tilde{\rho}_o(\tilde{J}). \tag{8}$$

Let us now calculate $\langle H \rangle$ in the original variables J, α. We then have by using equation (5) and the fact that Π is self-adjoint,

$$\langle H \rangle = \frac{1}{(2\pi)^N} \int dJ \int d\alpha H(J, \alpha) \rho(J, \alpha)$$

$$= \frac{1}{(2\pi)^N} \int dJ \int d\alpha H(J, \alpha) \Pi \rho(J, \alpha). \tag{9}$$

This shows that we need only the projection $\rho = \Pi_\rho$ of ρ into the null space of H. By using equation (6), this leads to further simplification; relation (6) may be written more explicitly $I_n = \mathcal{C}_n I_o$, where a Fourier expansion in angle variables is introduced:

$$\langle H \rangle = \frac{1}{(2\pi)^N} \int dJ \int d\alpha \left[H_o(J) \rho_o(J) \right.$$

$$+ \sum_{n \neq 0} V_n \exp(in\,\alpha) \sum_{m \neq 0} \rho_m(J) \exp(in\,\alpha) \right]$$

$$= \int dJ \left[H_o \rho_o(J) + \sum_{n \neq 0} V_{-n} \rho_n \right]$$

$$= \int dJ \left(H_o + \sum V_{-n} \mathcal{C}_n \right) \rho_o(J). \tag{10}$$

Once we know the distribution $\rho_o(J)$ of the unperturbed action variables, we may therefore calculate the average value $\langle H \rangle$, but we have to associate with H the *operator* on the action variables

$$\mathcal{K} = H_o + \sum_{n \neq 0} V_{-n} \mathcal{C}_n. \tag{11}$$

To this operator corresponds the eigenvalue problem

$$\mathcal{K} \varphi_{\tilde{j}}(J) = \tilde{H}(\tilde{J}) \varphi_{\tilde{j}}(J). \tag{12}$$

The eigenvalues are precisely the values of the Hamiltonian as expressed in the "correct" action variables (see eq. [7]). Therefore, the solution of this problem is equivalent to the solution of the Hamilton-Jacobi equation. Moreover, the eigenfunctions are *eigenprobabilities* which give the distribution of J once J is given. The analogy to quantum mechanics is striking.

It is quite remarkable that the concepts of classical mechanics when combined with the uses of an ensemble ρ lead to a new formulation of dynamics. This method can be taken over practically without change to quantum mechanics (see I. Prigogine and Cl. George, Physica [1971, in press]). Of course, the Liouville operator $L\rho$ is now associated to the commutator $[H, f]$. The usual way of quantization consists in looking for special forms of the density ρ (pure states) such as

$$\rho = |\Psi\rangle\langle\Psi|. \tag{13}$$

One obtains then the Schrodinger equation for $|\Psi\rangle$ whence the usual H-eigenvalue problem which leads to the quantization with, as eigenfunctions, probability *amplitudes*. This procedure corresponds therefore to the factorization of the Hilbert-Schmidt space $\mathfrak{h}^{(2)}$ of density matrices ρ into an exterior *product* of two Hilbert spaces $\mathfrak{h}$ for the probability amplitudes:

$$\mathfrak{h}^{(2)} = \mathfrak{h} \otimes \mathfrak{h} \tag{14}$$

and to the setting up of an eigenvalue problem in $\mathfrak{h}$. On the contrary, in our new method $\mathfrak{h}^{(2)}$ is written as the sum of two subspaces, symbolically

$$\mathfrak{h}^{(2)} = \Pi + \hat{\Pi}\tfrac{1}{2} \tag{15}$$

and the eigenvalue problem is formulated in the Π space also. This leads then again to an eigenvalue problem of the form (12) with eigenprobabilities as eigenfunctions. Let us make two final remarks. First, the solution of the eigenvalue problem (12) corresponds to the elimination of "correlations" as induced by the creation operator $\mathcal{C}$ in equation (11). Therefore, both the Hamilton-Jacobi method and the Schrodinger-Heisenberg method of diagonalization of H acquire a new meaning as methods for the *elimination* of correlations. Second, one method may be extended to more complicated situations (corresponding roughly to nonseparable systems in classical dynamics).

In conclusion, one method seems to us to have a special methodological interest, for it bridges concepts of classical or quantum dynamics with the ensemble concept of statistical mechanics.

R. Balescu: I should like briefly to point out a very remarkable property of the time evolution process in a many-body system. (This remark is based on recent work of I. Prigogine, Cl. George, F. Henin, J. Wallenborn, L. Brenig, and myself.)

The distribution function $f(t)$ can be separated into two terms by means of operators π and $\hat{\pi} \equiv 1 - \pi$:

$$f(t) = \pi f(t) + \hat{\pi} f(t). \tag{1}$$

It can be shown, under fairly general conditions, that (a) π (and hence $\hat{\pi}$) is a projection operator:

$$\pi^2 = \pi; \tag{2}$$

(b) π commutes with the Liouville operator $\mathcal{L}$:

$$\pi \mathcal{L} = \mathcal{L} \pi, \tag{3}$$

The property (3) is quite remarkable. If $u(t) \equiv \exp(t\mathcal{L})$, we see that

$$\pi f(t) = \pi u(t) f(o) = u(t) \pi f(o). \tag{4}$$

Hence $\pi f(t)$ [and also $\hat{\pi} f(t)$] obeys the Liouville equation. This implies that the set of distribution functions $f(t)$ has been split into two subsets which are invariant under the motion. In other words, the representation $u(t)$ of the group of translations is reducible (even in presence of interactions!).

The formalism can be extended into a relativistic theory. In our formulation we begin with the construction of ten generalized Liouvillians, $\mathcal{L}_i$, which generate the infinitesimal transformations of the Poincaré group. The generator of translations is just the ordinary Liouvillian $\mathcal{L}_H \equiv \mathcal{L}$. The pairwise commutators of these operators satisfy the characteristic Lie algebra relations of the Poincaré group.

In a recent paper (R. Balescu and L. Brenig, to be published in Physica, 1970) we have been able to prove that the separation (1) is invariant under all the transformations of the Poincaré group. This is expressed by the commutation relations $\mathcal{L}_i \pi = \pi \mathcal{L}_i$, $i = 1, \ldots, 10$, valid for *all* the Liouvillian generators. In particular, this theorem implies that under a Lorentz transformation, the πf component transforms into a πf component, independently of $\hat{\pi} f$. This invariance property shows that the separation (1) expresses a deep intrinsic feature of the time evolution process.

We cannot discuss here the important connections of these concepts with the theory of kinetic equations. These questions have been treated in detail in the literature (see, e.g., R. Balescu and J. Wallenborn, to be published in Physica, 1970).

Dynamics in General Equilibrium Theory

I find myself something of an outsider in writing a paper like this; and for a mathematician to write a paper with no technical mathematics is a real (but very constructive) confrontation with the problem of communication. What I hope to achieve here is to convey some of my ideas on the relation between dynamics and the traditionally static, economic equilibrium theory.

Since I have been working in mathematical economics, I have been struck by the number of attacks on general equilibrium theory, on mathematical economics and even economic theory in general. Coming from a radical background, I have much sympathy for some of the arguments brought forth.

I first want to say a few words on the subject of mathematics in economics and economic theory in general. The role of theory per se hardly requires defense; theory can give a deeper understanding of any subject, subtle relations are seen, inconsistent ideas are exposed, new horizons are revealed.

A criticism commonly made of economic theory is its failure to make predictions of crises in the country or to anticipate correctly unemployment or inflation. One must be cautious in the social sciences about looking toward physics for answers. However, some comparisons with the physical sciences seem profitable in connection with the above criticism. In those sciences, where theory itself is in a far more advanced state, limitations can be

seen in a similar way. For example a given individual human body functions according to physical principles; however no physical scientist would predict a heart attack. The physical theory gives understanding of aspects of what goes on in the human body only under very idealized conditions. The physical theories eventually play some role in the education of medical doctors, who can then say some things, some times about a patient's susceptibility to a heart attack, preventive measures, and cures.

The economy of the world or even a nation is a very complex phenomenon, like a human body, involving a number of factors, both economic and political. It is no more reasonable to expect economic theorists to predict a nation's economic future than for a theoretical scientist to predict the future health of an individual.

Questions about the need for mathematics in economic theory have been raised. Indeed, the successes of mathematics in economics have not been nearly as impressive as in physics. Yet the notion of money and prices already introduces mathematics into economics; and the mathematics becomes deeper with the equation of equilibrium, supply equals demand. When one considers the equation, supply equals demand for several interdependent markets, the mathematical problem already takes on some sophistication.

What is special about "general equilibrium theory" as opposed to economic theory in general? For me the importance of general equilibrium theory lies in its traditions which are deeper than any other part of economic theory. These traditions, which of course derive from actual economic history, explain why equilibrium theory has played such a central role and possesses such depth in content. Many of the procedures and mathematical methods used in other parts of economics grew out of developments in general equilibrium theory. Also equilibrium theory plays an important role in communication within the economics profession. Since most economists are knowledgeable in equilibrium theory, they can understand new ideas more readily when presented in that context. Also, since equilibrium theory has been studied so much, new ideas introduced there show weaknesses and also strengths most quickly.

After all of this is said, equilibrium theory will eventually stand or fall, depending on its truth as an important idealization of actual economic systems or as a model with values of justice, of efficient distribution and of utilization of resources. As a normative theory, I find great merit in its decentralization features (Schumacher's popular book, "Small is Beautiful" expresses some of my sentiment on decentralization). There are also, without doubt, basic failings in the theory which are well-presented in the economic literature, and there are some weaknesses which I wish to discuss presently. In fact these problems can be seen and understood especially clearly due to the well-developed structure of the theory; and one can use the body of general equilibrium theory as a tool in developing alternate models. To me it would seem overly difficult to construct and communicate successfully any alternate economic theory without having first studied very thoroughly equilibrium theory.

I would like to give some reasons why I feel equilibrium theory is far from satisfactory. For one thing the theory has not successfully confronted the question, "How is equilibrium reached?" Dynamic considerations would seem necessary to resolve this problem. Another weakness is the reliance of the theory on long range optimization.

In the main model of equilibrium theory, say as presented in Gerard Debreu's *Theory of Value*, economic agents make one life-long decision, optimizing some value. With future dating of commodities, time has almost an artificial role. The model is reminiscent of John von Neumann's game theory. I like to make an analogy between "Theory of Value" and the game theoretic approach to chess. The possible strategies are laid out to each player in advance, paths in a game tree, or a set of moves, one move to each position that could possibly occur. Each player makes a single choice of strategy. The strategies are compared and the game is over. Of course, chess isn't played like this. And in a situation more complicated than chess, where life-long consumption plans replace strategies, I don't believe economic agents act that way either.

In fact even the very best chess players don't analyze very many moves and certainly don't make future commitments. Their experience together with the environment at the moment (the position), some rules of thumb and some other considerations lead to decisions on the playing board.

My personal economic decisions are of a similar nature, from buying a book to buying a house; from a decision to travel to decisions about my job. Between one economic decision and another, there has been a real passage of time, circumstances have changed, and the new decision takes place in this new environment.

Long-run optimization would be impractical, even if it were emotionally acceptable, because of barriers of complexity. Complexity keeps us from analyzing very far ahead. The amount of time involved in making a decision is an important factor, for a chess player or a purchaser. Dynamic models based on some kind of behavioral strategy could meet these objections.

Sometimes static theories pose paradoxes whose resolution lies in a dynamic perspective. Let me give an example from the theory of duopoly of the Cournot (or Nash) equilibrium solution. Under classical hypotheses on the profit functions, consider such an equilibrium $(r_1, \; r_2)$ where r_i is the rate of production of duopolist number i. Then agent 1's rate of production maximizes profit among all such rates with r_2 fixed. On the other hand the solution is unsatisfactory because there is another state, say, (r'_1, r'_2), where each duopolist with reduced production is taking a higher profit. In the actual dynamic world it is unlikely that the duopolists would stay at the Cournot solution, knowing that they would both be better off at nearby states.

It doesn't require explicit cooperation for these agents to move off the Cournot solution. In fact with flow of information and implicit threats in the context of a passage of time, one can argue that the duopolists will move to an optimal state from a Cournot state. But this resolution requires a real passage of time, that after each market move, the opposing duopolist has another opportunity to move. One can readily think of examples of duopoly where an increase of advertising is withheld by one agent knowing the other agent would match an increase and both would be worse off. James Friedman has written on this topic.

I feel that dynamics could also play a role in the resolution of Kenneth Arrow's paradox in the politics of social choice.

Politics and elections in particular are actually part of a dynamic process, balloting being just a stage. The process looked at as a whole involves a number of moves such as a candidate's speech, a political ad, revising a position on some issue, etc. After each action of a candidate, other candidates have the option of taking an action of their own; voters' opinions evolve. For this reason, I would think that a dynamical model of the political process would give much better perspective than a static model of simple voting. In relation to this, the work of G. Kramer comes to mind.

We return to the subject of equilibrium theory. The existence theory of the static approach is deeply rooted to the use of the mathematics of fixed point theory. Thus one step in the liberation from the static point of view would be to use a mathematics of a different kind. Furthermore, proofs of fixed point theorems traditionally use difficult ideas of algebraic topology, and this has obscured the economic phenomena underlying the existence of equilibria. Also the economic equilibrium problem presents itself most directly and with the most tradition not as a fixed point problem, but as an *equation*, supply equals demand. Mathematical economists have translated the problem of solving this equation into a fixed point problem.

I think it is fair to say that for the main existence problems in the theory of economic equilibria, one can now bypass the fixed point approach and attack the equations directly to give existence of solutions, with a simpler kind of mathematics and even mathematics with dynamic and algorithmic overtones. In the last part of the paper we elaborate on this point.

Behind my own work on the questions of dynamics in economics, lies certain foundational work in the equilibrium theory in terms of calculus. The early development of mathematical economics, including the 19th century and even up to World War II, was largely in terms of calculus; it was no doubt the influence of game theory, and associated fixed point theorems that gradually reduced the dependence on calculus. In *Theory of Value* in 1959, Debreu wrote of the work of von Neumann and Oskar Morgenstern which freed mathematical economics from its traditions of differential calculus and compromises with logic. He wrote of the radical change of mathematical tools from calculus to convexity and topology.

But then in his paper on a finite number of equilibria, Debreu returned to calculus tools; my own work, "Global Analysis and Economics," has been to try to systematize the use of calculus in equilibrium theory. This can be justified on several grounds. First, the theory is brought closer to the practice. With calculus, one has in the derivative a linear approximation. It is these linear conditions that are so basic to practical economic studies. Comparative statics depend on derivatives; the same is usually true for stability conditions; dynamic questions are more accessible via calculus. When general equilibrium theory is developed on calculus mathematics, not only is theory brought closer to practice, but greater unity is achieved. Furthermore, recent work on approximation by differentiable functions in economics gives further justification to the use of calculus.

Finally before moving to the constructive side of the question of dynamics in equilibrium theory, it is worth making a remark on the nature of goods. It seems from our experience that it is important especially in modeling dynamics to put goods into two ideal classes, either completely perishable or completely durable. The theory seems different for the two kinds. For example, Walras equilibrium seems suited to the perishable, continually endowed class of goods, while for durable goods, the kind of equilibrium found in the fundamental theorem of welfare economics seems more appropriate. We hope that the rest of the paper makes this point clearer.

We discuss now the results of our paper, "Exchange Processes with Price Adjustment." This is a model of a market of durable goods; a particular example from personal experience of such a market is a weekend "mineral bourse" where agents with minerals and/or money meet to exchange, buy, and sell fine mineral specimens for collectors. Here one sees a truly dynamic process of exchange and price adjustments which converges to an equilibrium toward the end of the weekend of the "bourse."

An early work in some of the same spirit was done in 1962 by Frank H. Hahn, Hirofumi Uzawa, and Takashi Negishi. This approach is called a "nontatonment," because in place of the Walrasian Tatonment, a sequence of actual trades takes place over time. On the other hand, our model differs from the 1962 work in that it is nondeterministic, the dynamics proceeds without an ordinary differential equation, long-run optimization is dropped, and a large body of examples is constructed.

The main result of our paper can be stated as follows:

"In a pure exchange economy, an exchange price adjustment process, responsive to transaction costs, and which doesn't stop unless forced to by market conditions, converges to a price equilibrium. There exists such processes starting from any state of any (pure exchange) economy."

In the paper, mathematical content is given to all of the phrases used here, and the result is proved. Here we give a brief explanation of some of the terms used.

A "state" of an economy means a set of

data characterizing the economy at a given time. Our use of the word "state" is akin to its use in physics. For example in a pure exchange economy, a state will consist of an allocation of the resources, or equivalently the set of goods of each agent and a price system.

A state in general will change over time, e.g., by exchange and price adjustments. Thus a "process" as used in the main result, means a passage in time of a state. Or equivalently, a process is a path (over time) in the space of states of an economy.

This process, to qualify as an *exchange* process, must satisfy economically justifiable conditions for exchange which are embodied in the following axiom. The exchange axiom for the process asserts that: (a) the total resources of the economy are constant (there is no production); (b) exchange takes place at current prices; (c) an exchange increases satisfaction of the participating agents; and (d) some exchange will take place provided that it is possible consistent with (a), (b), and (c).

For the process to qualify as a price adjustment process (as in the quoted main result), we demand that it satisfy a price adjustment axiom defined in terms of a short run version of demand. A usual excess demand approach requires long-run optimization for the agents while our spirit is closer to that of behavioral strategies. At given prices and goods possessed one defines the infinitesimal demand of an agent to be the direction his preferences take him when restricted to his budget set.

The price adjustment axiom asserts that prices adjust in the direction of some weighting of the infinitesimal demands of all the agents.

A Walrasian price equilibrium depends on the traders' endowments. Thus if one allows a real passage of time, say an actual exchange to take place, and several such, this initial endowment becomes for-

gotten. Thus if one allows a "nontatonment" kind of time passage, one must replace a Walrasian price equilibrium by a different notion of price equilibrium.

As stated in our main result above, a price equilibrium is a feasible allocation and price system where, for each agent, satisfaction is maximized on his budget set defined relative to his wealth at equilibrium. Equivalently, a price equilibrium is an optimal allocation together with a supporting price system, as studied in the correspondence of the fundamental theorem of welfare economics.

A detailed mechanism of price setting and transactions is not developed, but it seems likely that the model is consistent with doing this explication.

Next we discuss some problems and results on the classical Walrasian model from the point of view of dynamics and algorithms. We prefer an alternate, well-known interpretation of the Walras model to that given in Debreu's *Theory of Value*.

Suppose that the goods are perishable, with labor a main example. One might envision a situation where each day an agent starts his economic activity with a fixed endowment of labor, or fish which won't keep. The next day he will have a new endowment of the same, but none left from the day before.

The consumption variables are the amounts of commodities consumed each day, of an agent. Thus both the endowment and consumption bundles in commodity space will be interpreted as the *rates* of endowment (fixed over time) and consumption respectively. A completely satisfactory dynamics (which isn't available) for this problem would construct and analyze paths over time in the space of states, that is commodity vectors for each agent and price systems (or sets of price systems). These paths should obey economically justifiable axioms of exchange and price adjustment, and probably should lead, at least under some economic conditions to a Walrasian equilibrium, starting from an endowment allocation and any price system. At the most satisfactory level, these paths should be given interpretations in terms of individual agent's actions in price offerings and purchases. In my view, a behavioral strategy for agents would be more desirable than decisions based on long run optimization.

Martin Shubik and I have worked on the problem in this setting without any definite success. On the other hand, it seems as if we might eventually obtain such a model with convergence provided a condition such as gross substitutes is satisfied. I should emphasize that I am not talking about any "tatonment" in this kind of dynamic, but rather an actual process, where agents are adjusting their goals, consumptions, prices over time to arrive at balanced budgets where the value of the rate of consumption equals the value of the rate of the fixed endowment for each agent.

I would like to turn now to some work carried out in my article "A Convergent Process of Price Adjustment and Global Newton Methods," which has more success on the mathematical side of the above problem.

One way of looking at this work is to first alter the Scarf Algorithm from finding fixed points to solving a system of equations, especially the system, supply equals demand in many variables. We define an ordinary differential equation, called a "global Newton," which is a version of (the altered) Scarf's Algorithm. Under rather general hypotheses (comparable to those needed in the execution of Scarf's Algorithm), solutions of the global Newton converge to the set of solutions of the original system (e.g., supply equals demand). Combining this fact with methods of numerical analysis,

one obtains a different but analogous algorithm to that of Scarf. Morris Hirsch and I have implemented this effectively on a computer and are developing the algorithm from a numerical analysis point of view. It applies to systems of n nonlinear equations in n variables, without hypotheses on the system of nonzero Jacobian, convexity, or monotonicity.

Let $z(p)$ be the excess demand as a function of prices $p = (p_1 \ldots p_l)$ so that p^* is an equilibrium if $z(p^*) = 0$. Then the global Newton takes the form

$$(1) \qquad Dz(p) \frac{dp}{dt} = -\lambda z(p),$$

$$\text{sign } \lambda = \text{sign Determinant } Dz(p)$$

Here $Dz(p)$ is the matrix of first partial derivatives of z. If one takes $\lambda = 1$ and uses Euler's method of discrete approximation to (1) then one obtains Newton's method for solving $z(p) = 0$. Using equation (1), one can obtain a proof of the existence of economic equilibrium without using fixed point theorems or algebraic topology.

Consider the problem of representing a process of price adjustments by (1). Recall that the classical "tatonment" process has an embodiment in the equation

$$(2) \qquad \frac{dp}{dt} = z(p)$$

Arrow, Leonid Hurwicz and Herbert Block have shown that solutions of (2) converge to economic equilibrium under hypotheses on z such as gross substitutes. These hypotheses are substantial and are strong enough to imply the existence of a unique equilibrium. On the other hand Scarf subsequently showed that under classical properties on preference relations, almost all solutions of (2) could oscillate for all time.

Now (1) can be considered as a modification of (2), which involves more subtle intermarket relations and which will converge when (2) doesn't. In particular (1) will converge in the Scarf example.

Generally speaking, (1) converges starting from almost any initial price system on the boundary of the price simplex. One could also formulate the equation in quantity space and look for an interpretation of the process in terms of budget balancing actions on the part of the agents. The interested reader could pursue these topics further in the papers cited.

REFERENCES

K. **Arrow**, *Social Choice and Individual Values*, New Haven 1963.

————— and L. **Hurwicz**, "The Stability of the Competitive Equilibrium I," *Econometrica*, 1958, *26*, 522–52.

—————, **Block** and L. **Hurwicz**, "The Sta-

bility of Competitive Equilibrium II," *Econometrica*, 1959, *27*, 82–109.

G. **Debreu**, *Theory of Value*, New York 1959.

—————, "Economies with a Finite Set of Equilibria," *Econometrica*, 1970, *38*, 387–92.

J. **Friedman**, "A Non-Cooperative Equilibrium for Super Games," *Rev. Econ. Studies*, 1971, *113*, 1–12.

H. **Scarf**, "Some Examples of Global Instability of the Competitive Equilibrium," *Int. Econ. Rev.*, 1960, *1*, 157–72.

—————, *The Computation of Economic Equilibria*, New Haven 1973.

S. **Smale**, "Global Analysis and Economics, IIA–VI," *J. Math. Econ.*, 1974–75, *1*, 1–14, 107–17, 119–27, 213–21.

—————, "A Convergent Process of Price Adjustment and Global Newton Methods," preprint, Berkeley.

—————, "Exchange Processes with Price Adjustment," preprint, Berkeley.

SOME DYNAMICAL QUESTIONS
IN MATHEMATICAL ECONOMICS

RESUME

Cette courte note met à jour mon article de l'American Economic Review [3].
Un thème particulier de cet article est développé ici, à savoir le lien entre la nature des
biens et la notion d'équilibre à retenir.

Let me start by posing what I like to call "the fundamental problem
of equilibrium theory" : *how is economic equilibrium attained* ? A dual
question more commonly raised is : *why is economic equilibrium stable ?*
Behind these questions lie the problem of modeling economic processes
and introducing dynamics into equilibrium theory. A successful attack here
would give greater validity to equilibrium theory. It may be however that a
resolution of this fundamental problem will require a recasting of the foundations of equilibrium theory. One might well keep in mind some historical
perspective from physics, making an analogy between Walrasian equilibrium
theory and Newtonian mechanics.

How did Relativity Theory respect classical mechanics ? For one thing
Einstein worked from a very deep understanding of the Newtonian theory.
Another point to remember is that while Relativity Theory lies in contra-
diction to Newtonian theory, even after Einstein, classical mechanics remains
central to physics. I can well imagine that a revolution in economic theory
could take place over the question of dynamics, which would both restruc-
ture the foundations of Walras and leave the classical theory playing a central
role.

In the direction of attacking this fundamental problem, it seems to me
important to idealize economic goods into two extreme classes. One one
side are the durable goods, and on the other, the perishable, renewable goods
with especially labor as an example. To each of these two classes of goods,
one can let correspond two basic branches of equilibrium theory. As an illus-
tration, Debreu's "Theory of Value" [1] has two substantive chapters,

Chapter 5 on the existence of (Walras) equilibria and Chapter 6 on "the fundamental theorem of welfare economics". I believe that one can associate the durable goods most naturally to the models in welfare economics and the renewable goods to the Walras equilibrium theory.

To see these things, it is useful to explicate the conditions for equilibrium. Assume classical (differentiable version) hypotheses on preferences for example as in [2]. Let there be l commodities, m agents in a pure exchange economy. Let price systems be denoted by $p = (p_1, \ldots, p_l)$ each $p_i \geqslant 0$, with $\Sigma p_i^2 = 1$. The endowment e_i of the ith agent will be a vector in $R = \{(e^1, \ldots, e^l)_i e^j \geqslant 0\}$ and an allocation will be an m-tuple

$$x = (x_1, \ldots, x_m),$$

with each x_i in R^l. The preference of agent i is supposed to be represented by a utility function $u_i : R_+^l \to R$.

A pair (x, p) consisting of an allocation and a price system is a Walras equilibrium if these equations are satisfied:

(1) The gradient, grad $u_i(x_i)$ equals λp for some $\lambda > 0$, each $i = 1, \ldots, m$. This is a necessary condition for x_i to maximize satisfaction for agent i.

(2) $\Sigma x_i = \Sigma e_i$. This is a total resource condition on the allocation x. In other terms, x is attainable or even "supply equals demand".

(3) $p \cdot x_i = p \cdot e_i$, $i = 1, \ldots, m$. These dot products give the values and this is a budget condition.

The first two equations by themselves describe the kind of equilibrium used in welfare economics (e.g. Debreu's Chapter VI).

Returning to problem of dynamics, observe that if one is trading a non-tatonment situation with durable goods (or stocks), then the endowment allocation, after some trades will lose its effect and therefore play no role in any equilibrium attained (see [4]). Thus the notion of equilibrium which is relevant is not that of Walras but that of welfare economics.

While in the durable goods market, a commodity vector x_i in R_+^l is interpreted as a stock of goods, in renewable goods models, a point in commodity space is more naturally interpreted as a rate (or flow) of endowments or consumptions.

Thus in a model where the endowment of goods is being renewed continually, the endowments e_i should play a role in the equilibrium attained and therefore, a Walras equilibrium defined by the full set of equations (1) - (3) is most reasonable.

Perhaps the non-tatonment theory initiated by Hahn, Negishi, Uzawa has developed to handle the dynamics of durable goods of pure exchange in principle. On the other hand, clearly there is no satisfactory model for dynamics of renewable goods and Walras equilibria.

REFERENCES

[1] DEBREU G. – Theory of Value, New York 1959.

[2] SMALE S. – "Global Analysis and Economics VI", Jour. *Math. Econ.*, 3, (1976), 1-4.

[3] SMALE S. – "Dynamics in General Equilibrium Theory". *Amer. Econ. Rev.*, 66, (1976), 288-294.

[4] SMALE S. – "Exchange Processes with price adjustment", (to appear, Journ. Math. Econ.).

DISCUSSION

The discussant, Egbert Dierker, points out that, in his opinion, not only the case of perishable but also that of durable goods exhibits a Walrasian character, since the distribution of initial endowments is important for the final outcome. Smale answers that it is useful to study the pure laboratory cases first. Gabszewicz points out that the use derived from a durable good can be treated as a flow. Smale answers that the market for minerals or for houses cannot naturally be described in terms of flows. Bliss asks whether Smale distinction is appropriate. Smale answers that it is essential to know how to deal with labor, a purely perishable good. Production may then play the role of bringing both, perishable and durable goods, together into one model. The problem, however, is how to put stocks and flows into the same model. Harsanyi remarks that the distinction between durable and perishable goods is not that between a tatonnement and a non-tatonnement situation. The distinction rather lies in the fact that resale is possible in the first case but not in the latter.

Fuchs supports Smale's distinction and remarks that the case of perishable goods can be treated by the theory of temporary equilibria. The discussant points that in most models of price formation expectations about future prices play a major role. He asks to what extent agents anticipate price changes in Smale's model. Smale answers that his model had to be altered if individual expectations of price variations are to be taken explicitly into account.

Gabszewicz further remarks that the question of how to connect a given state with a Pareto optimum in the Malinvaud-Drèze-de la Vallée Poussin model is closely related to Smale's treatment of the durable goods case. Champsaur and Cornet relate Smale's process to Malinvaud's. But the latter process does not converge in finite time. Smale remarks that it is important how equations are defined near equilibria. Fuchs points out that an interest of Smale's model is that the Pareto set is reached in finite time, so one may minimize the time necessary to reach the Pareto set from the initial state. Smale answers that he originally considered processes responsive to time cost.

Guesnerie asks whether it is possible to reach any individually rational Pareto optimum in the durable goods case. Smale answers that it is likely that one can reach a subset of the Pareto surface of full dimension. A result of this kind has been shown by Schecter in a similar model without price adjustment. Kirman explains that in Smale's process an individual may continuously make losses because expectation about price changes are neglected. Smale says that the conditions characterizing his process require an essential change if one wants to handle this problem. Last Selten remarks that a behavioral point of view may be more appropriate than the requirement of full maximization. In order to relate a theory to experiments is should be put in a discrete framework. Smale answers that is should be possible reformulate his theory in a discrete set-up.

BULLETIN OF THE
AMERICAN MATHEMATICAL SOCIETY
Volume 83. Numbe. 4, July 1977

Global variational analysis: *Weierstrass integrals on a Riemannian manifold*, by
  Marston Morse, Mathematical Notes, Princeton University Press, Princeton, New Jersey, 1976, ix + 255 pp., $6.50.

The first thing that comes to mind in reviewing a new book by Marston
Morse on the calculus of variations is that he wrote a book, *The calculus of
variations in the large*, forty years ago. The early book gave the foundations of
what is now called Morse theory. The publication of a new book by Morse on
the same subject presents an occasion to give some personal perspectives on
how this mathematics has developed in the last few decades. I say "personal
perspectives" and indeed, I, myself, have been involved in, and inspired by,
Morse's mathematics. For example, three of my papers contain the word
Morse in the title. Another mathematician much influenced by Morse, Raoul
Bott, was my adviser, and even work of Morse (but not variational theory)
suggested to Bott the thesis problem he gave me (leading eventually to my
work in immersion theory).

Another factor in writing a review like this is that, today, global analysis is
very much alive, both in mathematics and other disciplines. It may give us
some perspective to trace the development of one of the main roots of the
subject.

Let us see what Morse, in 1934, had to say about global analysis (he used
the word macro-analysis, then). I quote the full first paragraph of the
Foreword of his book.

"For several years the research of the writer has been oriented by a
conception of what might be termed macro-analysis. It seems probable to the
author that many of the objectively important problems in mathematical
physics, geometry, and analysis cannot be solved without radical additions to

the methods of what is now strictly regarded as pure analysis. Any problem which is nonlinear in character, which involves more than one coordinate system or more than one variable, or whose structure is initially defined in the large, is likely to require considerations of topology and group theory in order to arrive at its meaning and its solution. In the solution of such problems classical analysis will frequently appear as an instrument in the small, integrated over the whole problem with the aid of group theory or topology. Such conceptions are not due to the author. It will be sufficient to say that Henri Poincaré was among the first to have a conscious theory of macro-analysis, and of all mathematicians was doubtless the one who most effectively put such a theory into practice."

Note Morse's acknowledgment of the role of Poincaré in this subject. Already in 1885, Poincaré knew of Morse inequalities for a surface.

Note also how Morse feels that problems (of analysis) nonlinear in character are likely to require considerations of topology and group theory. I would like to echo this point, which even today, forty years later, has still not been digested by some analysts, analysts, for example, who are not willing to relinquish their linear spaces and linear space methods to confront nonlinear problems.

Suspicions of geometry and the uses of geometry in analysis have indeed deep roots. Even G. D. Birkhoff wrote in 1938, in *Fifty years of American mathematics*, of his " . . . disturbing secret fear that geometry may ultimately turn out to be no more than the glittering intuitional trappings of analysis." He used the word geometry to include topology or "analysis situs" as it was called then.

The simplest case of Morse theory is just the phenomenon that a differentiable function on an interval with two local minima must have a local maximum between the local minima. This "minimax principle", while very old, received a big push by G. D. Birkhoff in 1917. It is interesting to see what Morse had to say about some of the origins of the global calculus of variations. In his obituary of Birkhoff, he wrote (on the minimax principle):

"In Birkhoff's applications this principle reduces to an existence theorem for critical points of an analytic function $F(x)$ of n-variables. If one supposes for the sake of definiteness that F is defined over a regular, compact, analytic manifold, then, suitably counted, there exist at least $R_1 + M_0 - 1$ generalized saddle points, where R_1 is the linear connectivity of the manifold and M_0 the number of points (supposed isolated) of relative minima of F. In similar or related forms this principle was known and applied by Poincaré, Maxwell, and Kronecker, and has an origin even more remote in the past. Birkhoff's bold step was to conceive of its application to functions of curves such as the integral J. He applied it in the billiard ball problem [26] (motion on a convex table) and to obtain closed geodesics on a convex surface."

Thus, one has here a relation between the analysis, saddle points or geodesics, and the topology, "linear connectivities", or more generally, Betti numbers.

Morse's central contribution was to take these ideas and apply them systematically to a functional on the "manifold" of curves joining two points to deduce the existence of extremals, especially geodesics. In doing so, he developed relations between the critical points and the topology for any (smooth, nondegenerate) function on a compact manifold. In particular, it was a great accomplishment of Morse, in the years 1925–1930, to have given a global geometric abstract base for the variational calculus. To that base we proceed.

A fundamental insight here deals with the problem: How does the topology change as a function passes a nondegenerate critical point? To explicate matters, let $J: \Omega \to R$ be a C^∞ real valued function defined on a compact manifold, Ω. We will suppose that all maps are C^∞ and use the symbol Ω for a manifold for reasons that will become more natural later. A point x in Ω is called a *critical point* of J if it has zero derivative, i.e., if $DJ(x) = 0$. In that case the second derivative $D^2 J(x)$ is an invariantly defined bilinear symmetric form H_x on the tangent space T_x of vectors tangent to Ω at x. This form is called the *Hessian* and plays an important role in Morse theory.

Recall that for any bilinear symmetric form H on a linear space E, the *index* is the maximum dimension of a subspace on which H is negative definite. The *nullity* of H is the dimension of the *null* space, i.e., the set of v in E such that $H(u,v) = 0$ for all u in E. Then H is *nondegenerate* if its nullity is zero. If H is nondegenerate and E is R^n, then there are linear coordinates u on R^n such that

$$H(u,u) = - \sum_{i=1}^{k} u_i^2 + \sum_{i=k+1}^{n} u_i^2 .$$

Here k is the index.

All of these definitions pass over to a critical point. Thus the *index* of a critical point is the index of its Hessian; the critical point is called *nondegenerate* if its Hessian is nondegenerate, etc. A nondegenerate critical point is necessarily isolated. One knows more.

The *Morse lemma* asserts that if a critical point x of a function J is nondegnerate, then there are coordinates u near x to make J quadratic, i.e., so that $J(u) = H_x(u,u)$. My own belief is that the Morse lemma, while nice to know, is not vital, and in fact Morse theory develops more naturally, more conceptually, without it. To understand the topology of a function on a manifold, one uses a Riemannian metric to define gradient lines of the function. The choice of coordinates in the Morse lemma will not respect this metric. On the other hand, simply Taylor's formula already gives sufficient local information to do the "handle attaching".

Perhaps the preceding remark will become clearer as we proceed with our story of what happens to the topology as a nondegenerate critical point is passed. For any real number a let J_a be the set of all x in Ω with $J(x) \leqslant a$. If there is no critical value in the interval $[b,c]$, then $J^{-1}[b,c]$ is differentiably isomorphic to a product, $J^{-1}(b) \times [b,c]$. (A critical value is the value of a critical point.)

Now suppose there is exactly one critical point x^* with value c, and that x^* is nondegenerate with index k. How is the topology of $J_{c+\varepsilon}$ related to that of $J_{c-\varepsilon}$ for small enough ε? The fundamental result is that $J_{c+\varepsilon}$ is $J_{c-\varepsilon}$ together with a cell (or "handle") of dimension k attached. This handle attaching statement has three versions, which relate to three periods in the development of Morse's critical point theory; and the last two relate to the application of this theory to problems of topology.

The first of these versions is on the homology level and states the relative homology (over the rationals) result:

$$H_i(J_{c+\varepsilon}, J_{c-\varepsilon}) = 0 \quad \text{if } i \neq k,$$

$$\dim H_k(J_{c+\varepsilon}, J_{c-\varepsilon}) = 1.$$

Recall k is the index of the critical point.

This result is central in Morse's 1934 book and is used to deduce the existence of critical points. These critical points correspond to solutions of variational problems, in particular, to geodesics as we shall see later. Thus it is used to make a passage from topology to analysis and geometry.

The second version is a sharpening, explicated by Bott in 1959, which puts the theorem on a homotopy level. This result was used by Bott to study the homotopy of Lie groups. In particular, he obtained the first proof of the Bott periodicity theorems this way. Specifically, the stable homotopy groups of the unitary group U and the orthogonal group O satisfy

$$\pi_i(U) = \pi_{i+2}(U), \qquad \pi_i(O) = \pi_{i+8}(O) \quad \text{all } i.$$

Here for example one may think of U as the union of $U(n)$ over $n = 1, 2, 3,$ ... and then π_i is the ordinary homotopy group. This result was the starting point of K-theory.

Bott's version of handle attaching is the statement that $J_{c+\varepsilon}$ is homotopically equivalent to $J_{c-\varepsilon}$ with a cell D^k of dimension k attached by a homeomorphism from the boundary ∂D^k of the cell into the level surface $J^{-1}(c - \varepsilon)$. The homotopy statement yields the homology statement as a corollary.

The third version of handle attaching is on the diffeomorphism level which I believe I was the first to explicate. In fact, this was one very important ingredient in my solution of the "higher dimensional Poincaré conjecture", work on "handle body theory", and the structure of manifolds.

To work on the level of differentiable isomorphism, one must thicken D^k to bring the dimension up to the dimension of the manifold, say n. Thus, let a k-handle be $D^{n-k} \times D^k$. Then the strongest version of handle attaching asserts that $J_{c+\varepsilon}$ is diffeomorphic to $J_{c-\varepsilon}$ with $D^{n-k} \times D^k$ attached by an imbedding of $D^{n-k} \times \partial D^k$ into $J^{-1}(c - \varepsilon)$. The attaching process involves a smoothing at the corners. This statement yields the homotopy version as a corollary.

For my favorite proof of these theorems, one supposes that the manifold has a Riemannian metric (by imposition, if necessary) and uses the flow defined by the negative of the gradient of J. At the critical point x^*, the linearized flow has two invariant subspaces in the tangent space T_{x^*}. One of these is contracting under the flow, say E^{n-k}, and one is expanding, say E^k (with the dimension of E^k equal to k, the index of x^*). Use these spaces to define a local product structure in the manifold near x^*; then take small disks D^{n-k}, D^k about the origins in E^{n-k}, E^k, respectively. Using Taylor's formula to expand J about x^*, one can show that the flow on the boundary of $D^{n-k} \times D^k$ has the requisite properties to give the attaching statements. The same proof works at the homotopy and diffeomorphism levels, but requires slightly more checking in the latter case.

The passage from the homology version of handle attaching to the Morse inequalities proceeds most simply via the exact homology sequence of a pair.

Suppose that there are real numbers, $c_0 < c_1 < c_2 < \cdots < c_m$, so that each interval (c_j, c_{j+1}) contains the value of exactly one critical point of $J: \Omega \to R$ and all the critical values are in such intervals. For each j, one has the exact homology sequence of vector spaces over the rational numbers, writing J_j for J_{c_j},

$$\to H_i(J_{j-1}) \to H_i(J_j) \to H_i(J_j, J_{j-1}) \to H_{i-1}(J_{j-1}) \to .$$

From linear algebra, summing from $i = 0$ to k yields for each k:

$$\sum_{i=0}^{k} (-1)^{k-i} \dim H_i(J_{j-1}) - \sum_{i=0}^{k} (-1)^{k-i} \dim H_i(J_j)$$

$$+ \sum_{i=0}^{k} (-1)^{k-i} \dim H_i(J_j, J_{j-1}) \geqslant 0.$$

Summing over j and evaluating by the handle attaching theorem gives

$$- \sum_{i=0}^{k} (-1)^{k-i} \dim H_i(J_m) + \sum_{i=0}^{k} (-1)^{k-i} M_i \geqslant 0$$

where M_i is the *Morse type number* or the number of critical points of index i. Let B_i be the ith Betti number, or $\dim H_i(\Omega)$, and since $J_m = \Omega$, we have

Morse inequalities. $\sum_{i=0}^{k} (-1)^{k-i} M_i \geqslant \sum_{i=0}^{k} (-1)^{k-i} B_i$, each $k = 0, 1, 2,$ ....

By adding these inequalities for k and $k - 1$, we get

Simple Morse inequalities. $M_k \geqslant B_k$, $k = 0, 1, 2, \ldots$.

In the preceding discussion we have assumed that different critical points took different values; a minor extension in the handle attaching argument can remove that hypothesis. We have also assumed that all of the critical points of J were nondegenerate. Such functions are called *Morse functions*.

How general are Morse functions? Morse has in his 1934 book a prototype of the theorem that Morse functions form an open and dense set among all C^∞ functions. This result is now centrally imbedded in transversality theory and the theory of singularities of maps à la Whitney, Thom and Mather.

121

Sard's theorem that with enough differentiability the values of critical points form a set of measure zero is basic in this development and it is no accident that Sard was a student of Morse.

It is important to remark that a complement to Morse's work was developed, already by 1930, by Lusternik and Schnirelmann in the Soviet Union. These mathematicians gave a global existence theory for critical points without making use of nondegeneracy hypotheses.

What we have described above is still in the finite dimensional and abstract realm, thus twice removed from Morse's contributions to the study of geodesics. But before we turn to that study, we give a couple of ways this finite dimensional theory has made itself felt in analysis in the domain of our own experience.

One way is in dynamical systems, where starting from a Morse function J on a Riemannian manifold, one obtains a differential equation with a rather simple structure. The negative of the gradient vector field of J has the property that along solutions (of $dx/dt = -\operatorname{grad} J(x)$), J is never increasing. Thus the dynamical behavior permits no nontrivial periodicity or recurrence. Furthermore, the zeroes of this differential equation, coming from a nondegenerate critical point, possess a certain local robust character. If one adds a second condition of transversality, that the asymptotic sets (the "stable and unstable manifolds") of these zeroes intersect transversally, one can obtain such properties globally. Via this route is the result I obtained with Jacob Palis, that every compact manifold supports a structurally stable dynamical system (so that the "phase portrait" or qualitative structure persists under perturbations). Moreover, it was the interplay between dynamical problems and topology that helped lead me to the handlebody results mentioned earlier.

In this vein, it is worth remarking that the catastrophes of Thom deal with bifurcations of gradient dynamical systems.

A second example is in celestial mechanics where one can show that the relative equilibria in the Newtonian n-body problem correspond to critical points of the Newtonian potential function on complex projective space, properly interpreted. Morse theory of this function suggested to Julian Palmore the existence of new relative equilibria in the 4-body problem which he found.

Let us turn now to the variational theory of Morse, the ultimate object of the abstract theory previously discussed and the subject proper of the book under review. The prinicpal example is the global study of geodesics on a manifold. Let us see how this goes.

Let M be a Riemannian manifold. Recall that this equips each tangent space $T_x = T_x(M)$, x in M, with a norm written $\| \ \|_x$ or sometimes simply $\| \ \|$, defined by an inner product $(\ , \ )_x$ in T_x. One can define the length $l(\alpha)$ of a curve $\alpha: [a, b] \to M$ by

$$l(\alpha) = \int_a^b \|\dot{\alpha}(t)\| \, dt \quad \text{where } \dot{\alpha} = \frac{d\alpha}{dt}.$$

From this, define a metric on M by letting $d(p,q)$ equal the infimum of $l(\alpha)$ over all curves α joining p to q. This is indeed a metric. We will assume that M is complete for this metric.

On M, let points P, Q be given. Denote by $\Omega = \Omega_{P,Q}(M)$ the *loop space*, or set of all (C^∞) curves on M from P to Q. That is,

$$\Omega = \{\alpha: [0,1] \to M \mid \alpha(0) = P, \alpha(1) = Q\}.$$

The *energy* is the map $J: \Omega \to R$ given by $J(\alpha) = \int_0^1 \|\dot\alpha(t)\|^2\, dt$.

Let us look at the "first variation formulae" of this calculus of variations problem defined by J. The space of variations of α, or the "tangent space" of Ω at α, is the linear space defined by

$$\{\eta: [0,1] \to T(M) \mid \eta(t) \in T_{\alpha(t)}(M), \eta(0) = 0, \eta(1) = 0\}.$$

One can think of $T_\alpha(\Omega)$ as the space of vector fields along the curve α.

One can think of geodesics as being like "critical points" of $J: \Omega \to R$. It would be natural to define the derivative of J at $\alpha \in \Omega$, $DJ(\alpha): T_\alpha(\Omega) \to R$ as follows.

In case M is R^n, then the tangent bundle $T(M)$ is $M \times R^n$ and one can proceed by letting $F: T(M) \to R$ be defined by $F(x, \dot x) = \|\dot x\|_x$ for $\dot x \in T_x$. Then for $\eta \in T_\alpha(\Omega)$ let

$$DJ(\alpha)(\eta) = \int_0^1 \frac{\partial F}{\partial x}\eta + \frac{\partial F}{\partial \dot x}\dot\eta\, dt,$$

where $\partial F/\partial x = (\partial F/\partial x)(\alpha(t), \dot\alpha(t))$ is a linear map on R^n for each t. Then if $DJ(\alpha) = 0$,

$$DJ(\alpha)(\eta) = \int_0^1 \left(\frac{\partial F}{\partial x} - \frac{d}{dt}\frac{\partial F}{\partial \dot x}\right)\eta = 0 \quad \text{for all } \eta \in T_\alpha(\Omega).$$

So Euler's equation for α is satisfied or:

Euler's equation: $d/dt\, \partial F/\partial \dot x\, (\alpha(t), \dot\alpha(t)) - \partial F/\partial x\, (\alpha(t), \dot\alpha(t)) = 0$.

If M is not R^n, then one has the argument and equation valid in each coordinate chart, or one could use the covariant derivative.

Recall that a curve $\alpha: [a, b] \to M$ is a *geodesic* if it locally minimizes length. Then the above kind of derivation shows that α in Ω is a geodesic if and only if α is a solution of Euler's equation in each coordinate chart. Thus the geodesics are indeed like critical points of J.

The earliest part of our review motivates the question as to whether there are the concepts of index and nullity for a geodesic α. In the "second derivative" of J at α, as before, one can find an answer. In a coordinate chart of M, it is natural to write for $D^2J(\alpha)$, the second derivative of J at α, the symmetric bilinear form on $T_\alpha(\Omega)$ defined by

$$D^2J(\alpha)(\eta, \xi) = H_\alpha(\eta, \xi) = \int_0^1 F_{xx}(\eta, \xi) + F_{x\dot x}(\eta, \dot\xi) + F_{\dot x x}(\dot\eta, \xi) + F_{\dot x\dot x}(\dot\eta, \dot\xi)\, dt.$$

Here, $\eta, \xi \in T_\alpha(\Omega)$ and $F_{xx} = F_{xx}(\alpha(t), \dot{\alpha}(t))$ is the second partial derivative of F as a bilinear form on $T_{\alpha(t)}(M) = R^n$, etc.

One defines the *index* and *nullity* of α simply as the index and nullity of H_α on $T_\alpha(\Omega)$. Furthermore, α is *nondegenerate* if H_α is.

Define an inner product on $T_\alpha(\Omega)$ via that on M: i.e.,

$$(\eta, \xi)_\alpha = \int_0^1 (\eta(t), \xi(t))_{\alpha(t)} \, dt, \qquad \eta, \xi \in T_\alpha(\Omega).$$

Using integration by parts, one obtains that for all η, ξ in $T_\alpha(\Omega)$,

$$H_\alpha(\eta, \xi) = (L\eta, \xi)$$

where

$$L\eta = -\frac{d}{dt}(F_{\dot{x}\dot{x}} \dot{\eta} + F_{x\dot{x}} \eta) + F_{xx} \eta - F_{x\dot{x}} \dot{\eta}$$

is the *Jacobi* (linear) *differential operator*.

One may express L invariantly in terms of the Riemann curvature tensor.

Say that P and Q are *conjugate* along α if $L\eta = 0$ for some nonzero $\eta \in T_\alpha(\Omega_{P,Q})$. The *multiplicity* of this conjugacy is the dimension of the linear space of such η.

The *Morse index theorem* asserts that the index of α is equal to the number of points $\alpha(t)$, $0 < t < 1$, such that $\alpha(t)$ is conjugate to $\alpha(0)$ along α, counting conjugate points with multiplicity. I like to think of this result as belonging to the spectral theory of differential operators.

To prepare us for his main result, Morse shows that for prescribed P in M, if Q is excluded from a set of measure zero in M, then all the geodesics in $\Omega_{P,Q}$ will be nondegenerate. Thus for such Q, J is like the Morse functions defined earlier. We may call such a pair (P, Q) a *nondegenerate pair*.

Now we may state the following basic theorem of Morse in the calculus of variations.

THEOREM. *Let (P, Q) be a nondegenerate pair on a complete Riemannian manifold M. Let B_i denote the dimension of the homology group $H_i(\Omega_{P,Q}(M))$ (over the rationals) of the loop space. Let M_i denote the number of geodesics joining P to Q (in Ω) of index i. Then the B_i, M_i satisfy the Morse inequalities*

$$M_0 \geqslant B_0, \qquad M_1 - M_0 \geqslant B_1 - B_0, \quad etc.$$

as before.

Morse proves the theorem by taking a sequence of finite dimensional manifolds which approximate Ω and applying the earlier abstract theory. These approximating manifolds are manifolds of piecewise geodesic curves.

As a particular case of this theorem, Morse in his first book takes M to be homeomorphic to the n-sphere S^n. By applying the same Morse inequalities to the standard Riemannian n-sphere in R^{n+1}, he is able to compute the B_i, Betti numbers of the loop space, which are given by (say for $n > 2$), $B_i = 1$ for $i = 0$, $n - 1$, $2(n - 1)$, $3(n - 2)$, ... and zero otherwise. Therefore, he concludes that for an arbitrary Riemannian structure on S^n, the existence of geodesics of index 0, $n - 1$, $2(n - 1)$, ... joining P to Q.

Applications of this basic theorem to a more general class of manifolds were hampered by lack of knowledge of the homology of loop spaces. In fact, the breakthrough on this problem didn't come until 1951 with Serre's thesis. Using the Leray spectral sequence, Serre showed that $H_i(\Omega(M)) \neq 0$ for a sequence of i going to infinity for a compact manifold M with finite fundamental group. From the Morse theorem, this is enough to conclude that on any compact Riemannian manifold, any nondegenerate pair (P, Q) is joined by an infinite number of geodesics.

The history of the problem of *closed* geodesics on a manifold homeomorphic to a sphere S^m is one with fine achievements; it is also a subject over which many mathematicians have stumbled. For example, in the foreword of his 1934 book, when discussing Poincaré's work on existence of closed geodesics on a convex surface, Morse says the validity of Poincare's reasoning "has been questioned". Explicit objections are presented by Morse in his Chapter 9. These last two chapters of his book in fact are devoted to showing the existence of $m(m + 1)/2$ closed geodesics of a special kind on a Riemannian manifold homeomorphic to S^m. Yet this work of Morse is in error, as was pointed out by A. S. Svarc in 1960. Bott in 1954 gave a new proof which again was in error as Svarc noted.

Some successes in this line have been accepted. Lusternik and Schnirelmann (1929) showed the existence of three closed geodesics without self-intersection on a surface homeomorphic to S^2. G. D. Birkhoff in 1927 showed that a manifold homeomorphic to S^n had at least one closed geodesic.

In the last two decades, these questions have been pursued with very substantial success, especially by Soviet and German mathematicians. Klingenberg in a recently printed set of notes (Bonn, 1976) gives an account of the subject of closed geodesics complete with history. These notes include a proof of his newest theorem: On every simply connected manifold there exist infinitely many closed (prime) geodesics. (Let us hope ....)

At this point, we add that two expositions of Morse theory have probably been much more widely read than Morse's original treatise. These are the books of Seifert and Threlfall and of Milnor. Especially for those who have learned their mathematics in recent decades, Milnor's book is to be recommended.

The language and concepts in the calculus of variations since 1736 (Euler) have suggested that extremals could be thought of as critical points of the functional, length, area, energy, etc. Our review emphasizes this analogy. In fact, this analogy is actual. The geodesics *are* critical points of the energy. One can put a manifold structure on Ω in such a way that the energy J becomes a differentiable map on Ω and an abstract Morse theory on infinite dimensional manifolds can be developed which yields Morse's theorems for geodesics. This is what Palais and I did in 1962–1964. One obtains a unification of the first and second parts of the material of our review. For example, a geodesic is literally a critical point, the definition of Hessian and index of an abstract critical point apply directly to give these notions for geodesics as a special case. One needs no longer to approximate Ω by finite dimensional manifolds of broken geodesics. The abstract theory already applies to Ω.

Two contributions had made the way easier for us. First, Eells had put an infinite dimensional manifold structure (locally Hilbert or locally Banach) on certain function spaces in 1958. Secondly, Lang did the foundations of differential topology for Banach manifolds in 1962. Then Ralph Abraham, Palais, and I, in 1962, worked out systematically some theory of manifolds of function spaces and properties of manifolds of function spaces.

The idea for this way of doing Morse theory is to replace the compactness condition of Ω by a compactness condition on the map J, a condition which Palais and I called *Condition* C.

More precisely, let Ω be a complete Riemannian manifold, no longer compact or even finite dimensional, but defined as before with coordinate charts as open sets in Hilbert space. Let J be a (smooth) function on Ω which is bounded below, has nondegenerate critical points (of finite index for simplicity) and satisfies:

CONDITION C. If α_i in Ω is a sequence with $J(\alpha_i)$ bounded and such that $\|DJ(\alpha_i)\|$ tends to zero, then α_i has a convergent subsequence.

The result is that for such a function, the Morse inequalities are true, relating the type numbers defined by critical points of J and the Betti numbers of Ω. The proof is essentially that given above via handle attaching.

Now for the Morse theory of geodesics, one only has to show that the energy J on the loop space Ω satisfies the above properties, with an appropriate manifold structure on Ω. Palais and I used the Sobolev completion of Ω, with norm defined by L^2 and first derivatives in L^2. Actually it was Palais who wrote out the theory for this case in his article in *Topology* in 1963.

Some of the new results on closed geodesics mentioned earlier were proved using Condition C as above.

Recently, Tony Tromba seems to have found a drastic modification of Condition C, and developed a Morse theory for geodesics using infinite dimensional manifolds with a space of much smoother curves.

Now I would like to spend the last few words of this review on the global variational calculus for more than one independent variable. It was in fact just this problem that led me into infinite dimensional manifolds in the early sixties. It remains a problem; although very recently, especially through the work of Tromba and Karen Uhlenbeck, there seem to be signs of progress.

The most interesting case for more than one independent variable is minimal surfaces. In the theory of Plateau's problem, I had been intrigued by a result of Morse-Tompkins and Shiffman in 1939. Their theorem asserted that if a Jordan curve in R^3 spans two stable minimal surfaces, then it spans a third of unstable type. This was suggestive of a Morse theory for Plateau's problem.

In the sixties I tried without success to find such a theory, or to imbed the Morse-Tompkins-Shiffman result in a conceptual general setting. Tromba and Uhlenbeck may now have succeeded in initiating a development of calculus of variations in the large for more than one independent variable.

REFERENCES

1. G. D. Birkhoff, *Collected mathematical papers*, Vols. I, II, III, Amer. Math. Soc., Providence, R.I., 1950.
2. W. Klingenberg, *Lectures on closed geodesics*, 2nd rev. ed., Univ. of Bonn, 1976.
3. J. W. Milnor, *Morse theory*, Ann. of Math. Studies, no. 51, Princeton Univ. Press, Princeton, N.J., 1963. MR **29** #634.
4. M. Morse, *The calculus of variations in the large*, Amer. Math. Soc. Colloq. Publ., vol. 18, Amer. Math. Soc., Providence, R.I., 1934.
5. ——, *George David Birkhoff and his mathematical work*, Bull. Amer. Math. Soc. **52** (1946), 357–391; reprinted in Collected Mathematical Papers, Vol. I, Amer. Math. Soc., Providence, R.I., 1950, pp. xxiii–lvii.
6. H. Poincaré, *Sur les courbes définies par les équation differentielles*, part 3, Liouville J. (4) **1** (1885), 167–244.
7. H. Seifert and W. Threlfall, *Variationsrechnung im Grossen* (*Morsesche Theorie*), Teubner, Leipzig, 1938.

STEPHEN SMALE

BULLETIN OF THE
AMERICAN MATHEMATICAL SOCIETY
Volume 84, Number 6, November 1978
© American Mathematical Society 1978

Catastrophe theory: *Selected papers*, 1972–1977, by E. C. Zeeman, Addison-Wesley, London, Amsterdam, Ontario, Sydney, Tokyo, 1977, x + 675 pp., $26.50 (hard binding), $14.50 (paper binding).

For the general public, catastrophe theory (or CT) has become the biggest thing in mathematics. René Thom and Christopher Zeeman are the two leaders of this field. L'Express (October 14–30, 1974) asserts that the "new Newton" is French (i.e. Thom). An announcement of Zeeman's lecture at Northwestern University in the spring of 1977 contains a quote describing catastrophe theory as the most important development in mathematics since the invention of calculus 300 years ago. Newsweek has given similar comparisons. Zeeman juxtaposes Newton and Thom in the volume under review (briefly ZCT), p. 623. Thom writes " . . . CT is–quite likely–the first coherent attempt (since Aristotelian Logic) to give a theory on *analogy*." [p. 637, ZCT]. On the back cover of Thom's book, *Structural stability and morphogenesis* [English translation, Benjamin, 1975 or Thom's SSM], is the quote from the London Times review, "In one sense the only book with which it can be compared is Newton's *Principia*."

Recently however, the importance of CT has been sharply and publicly challenged by Hector Sussman and subsequently by Sussman and Raphael Zahler [*Catastrophe theory as applied to the social and biological sciences*: *A critique* to appear in Synthese]. A critical story on CT by Gina Kolata in "Science", April 15, 1977, is headed: *Catastrophe theory*: *The emperor has no clothes*. A front page story on the New York Times, November 19, 1977 focuses on the challenges to CT.

To write a review in this environment has a very personal side for me. On one hand my own work on dynamical systems is closely connected to the origins of CT. I have had a long and close personal and professional relationship with both Thom and Zeeman. More than 20 years ago I was discussing singularities of maps, transversality, and immersions with René Thom. Thom tried to interest me in an early draft of chapters of his book *Structural stability and morphogenesis* in 1966.

On the other hand I have remained skeptical and aloof from CT, perhaps due to my conservatism in science. While my colleagues and students were showing enthusiasm for CT, I gave critical lectures, one at the University of Chicago in 1974, one at the Aspen Institute of Physics in 1975. More recently I have been quoted negatively in the "Science" and New York Times references above. This is the first time I have written on the subject, and I should warn the reader of this negative bias, far from shared by many of my fellow mathematicians.

Some of the mathematics underlying CT, especially transversality, and singularities of maps, has played a constructive role in outside disciplines, and is destined to play an ever increasing role. On the other hand I feel that CT itself has limited substance, great pretension and that catastrophe theorists have created a false picture in the mathematical community and the public as to the power of CT to solve problems in the social and natural sciences. It is to the credit of Sussman and Zahler that they have seriously challenged this false picture.

On this matter of warning to the reader, there is also the problem of my "quoting out of context". In fact the quotes I have made above and will make below are often imbedded in more prosaic language, perhaps tempering the brief and sometimes sharp sounding statements I refer to.

At this point, I would like to acknowledge my debt to J. Guckenheimer, C. Zeeman, H. Sussman, and M. Hirsch among others for very useful conversations on the subject here. Besides the work of Thom and Zeeman, on Catastrophe Theory, I have also learned much from that of Sussman and Zahler.

Just what is Catastrophe Theory? Thom writes that CT " . . . has to be considered as a theory of general morphology" 633, ZCT and Zeeman "CT is a new mathematical method for describing the evolution of forms in nature." (1, ZCT).

However the new mathematics associated to CT really is contained in what is called elementary catastrophe theory (ECT) by Thom and Zeeman. Very briefly, ECT studies a smooth real valued map f as a function (often called a "potential function") of a state x and parameter μ. Here the state lies in some Euclidean space and μ also varies in a (usually low dimensional) Euclidean space. The problem is to find local canonical forms for "generic" f, using a smooth change of coordinates in the variables x, μ, $y = f(x, \mu)$. In this case, "generic" means for an open dense subset in a suitable function space. A local canonical form is defined in a neighborhood of a pair (x_0, μ_0) at which f is singular. ECT solves this problem when μ is in a low (≤ 5) dimensional space. In particular, when μ lies in a 4 dimensional space (e.g. space-time), Thom found seven canonical forms. We will shortly describe the case of the cusp catastrophe where x is in $\mathbf{R}$ and μ in $\mathbf{R}^2$.

Is CT much more than ECT? Here Thom and Zeeman differ. Thom writes " . . . that Christopher's criticisms arise basically from a strict dogmatic view of CT which he identifies with ECT . . . " 633, ZCT. Zeeman replies (same page) that his emphasis on ECT has been mainly because of its usefulness in applications. In fact, the book under review, although titled CT, deals almost completely with ECT (its mathematics and applications).

In my mind, when CT goes beyond ECT it loses pretty much any direct touch with mathematics. It is true that Thom refers to nonelementary CT as "generalized catastrophes, composed map catastrophes, G-invariant catastrophes" etc. 633, ZCT. But here the relationship of the mathematics to nature in the CT literature becomes tenuous and infrequent. For example in Thom (SSM), the words "generalized catastrophe" are used to describe situations where no mathematical model is proposed. More particularly under the picture of "feather buds on a chicken embryo", Thom writes "a generalized catastrophe and example of symmetry in biology" (Figure 21, Thom's SSM). He labels a picture of "the Crab nebula, the remains of the explosion of a supernova" with "a partially filament catastrophe in astrophysics" (Figure 17, Thom's SSM). Thom described the filament catastrophe as a certain "codimension two" type of generalized catastrophe, but the mathematics is not specified.

No doubt it is in this spirit Thom writes (p. 189 in *Structural stability, catastrophe theory and applied mathematics*, SIAM Rev., April, 1977) "The truth is that CT is not a mathematical theory, but a body of ideas, I dare say a state of mind."

In ECT, one can take the gradient of the "potential" function with respect to the state variable to obtain a dynamical system $x' = \text{grad} f_\mu(x)$ parameterized by μ, $f_\mu(x) = f(x, \mu)$. It is in this context that Thom introduced ECT in his book SSM, Chapter 5. He takes μ to be in $\mathbf{R}^4$ or space-time. Elementary catastrophes are defined there as the points μ of space-time for which the qualitative dynamics of $x' = \text{grad} f_\mu(x)$ changes in a neighborhood of the attractors (stable equilibria) of that dynamics. Generalized catastrophes are defined analagously where the gradient dynamical systems are replaced by a more general class of dynamical systems.

However the mathematical theorems that Thom states (later proved by Mather) actually give the classification of elementary catastrophes according to the theory of singularities of maps in contrast to the theory of dynamical systems. Guckenheimer subsequently pointed out (Bull. Amer. Math. Soc. **79** (1973), 878–890, and in *Bifurcation and catastrophe*, Dynamical Systems (ed., Peixoto) Academic Press, 1973) that the two classifications differ already when μ is in $\mathbf{R}^3$ (and certainly for μ in $\mathbf{R}^4$). I believe this weakens much of the scientific or philosophical foundations of CT which is a theory based on dynamical systems (Thom's SSM). If even the elementary catastrophes don't correspond to the dynamical classification, what of the generalized catastrophes where the mathematics in SSM deals in only a few known examples?

In fact ECT seems much closer to what economists call comparative statics than to dynamical systems.

One can sense a corresponding change of perspective in the work of the catastrophe theorists. As we have noted, the starting point of CT in Thom's SSM was the "bifurcations" of dynamical systems parameterized by space-time. In Zeeman's work, these parameters changed from "space-time" to "control parameters". Now in his SIAM article (cited above) Thom also speaks of control parameters and seems to de-emphasize the dynamical system foundations of CT, especially relative to space-time.

The cusp catastrophe is the most important example of a catastrophe and much if not most of the book under review (ZCT) centers around the cusp catastrophe and its applications. Zeeman with Isnard (ZCT, 329 in *Some models from catastrophe theory in the social sciences*) describes the canonical model as follows. In 3 dimensional space $\mathbf{R}^3$, variables (a, b, x), let M be the cubic surface defined by $x^3 = a + bx$. Here a, b are horizontal axes and are the controls; x is the vertical state variable. The fold curve F is where the vertical lines are tangent to M and is given by $3x^2 = b$. The projection of F onto the control space is called the bifurcation set B. The equation of B is $27a^2 = 4b^3$ and has a cusp at the origin. It is supposed that M is given by $\partial P / \partial x = 0$ where P is some (probability or potential) function $P: \mathbf{R}^3 \to \mathbf{R}$

and that a sociologically (or physically) meaningful subset G of M is the set of local maxima (or local minima) of P.

The state of the system stays on G and is controlled by the choice of (a, b). Thus when (a, b) crosses B, the state may be forced to make a discontinuous jump to remain on G.

This is all laid out clearly in the following Figure 330, ZCT from Zeeman and Isnard.

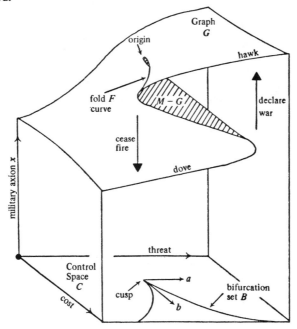

FIGURE 11*

In the model under discussion, a is a numerical representation of a threat, b the cost and x represents military action. The discontinuity, representing a jump in military action could be thought of as a declaration of war.

Thus what Zeeman and Isnard have given us is a model for the study of a nation deciding upon its level of action in some war. I consider this paper the most developed of Zeeman's papers on CT and the social sciences and he writes "And I believe that sociology may well be one of the first fields to feel the full impact of this new type of applied mathematics, . . . " (627, ZCT).

Sussman and Zahler have discussed this model in detail. Here I would like to focus on the question of its *justification* as a model of military decision

---

*Reprinted from the chapter entitled "Some models from catastrophe theory in the social sciences" by C. A. Isnard and E. C. Zeeman in *The Use of Models in the Social Sciences* (Social Issues in the Seventies Series) edited by Lyndhurst Collins, Tavistock Publications, London, copyright © 1976 Seminars Committee of the Faculty of Social Sciences of the University of Edinburgh, Scotland.

making; in this I believe the authors have failed. Their efforts in this direction lie on the mathematical side. Zeeman and Isnard write " . . . we shall introduce sociological hypotheses, and translate them into mathematics. The deep theorems of CT will enable us to synthesize the mathematics. We can then translate the synthesis back into sociological conclusions. It is not immediately apparent, without the use of the intervening mathematics, that the sociological hypotheses imply the sociological conclusions, and that is the purpose of using catastrophe theory." (314, ZCT).

The trouble is with the sociological hypotheses. Evidence for them, or for the model, should be given, for example from military history, studies in decision making, or sociological, political studies in general. There is much theory and data from history and social sciences relevant to the model of Zeeman and Isnard. None of this finds its way into the paper directly or indirectly save for brief references to Tolstoy on calculus in *War and peace* and Lorenz, *On aggression*.

Let us look at some of the sociological hypotheses that Zeeman and Isnard in fact do make. Summarized into mathematics, they express these hypotheses graphically by:

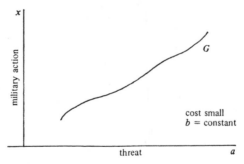

FIGURE 6, 316, ZCT

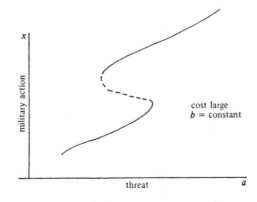

FIGURE 7, 317, ZCT

Zeeman and Isnard then ask how does Figure 6 evolve into Figure 7. "The main theorem of CT tells us that qualitatively there is only one way for this evolution to occur" [p. 318, ZCT], and they deduce Figure 11, cited above.

One trouble is the definition for "small" cost and "large" cost that Zeeman and Isnard use. They mean there exists b_0 such that "small b" means $b < b_0$ and "large b" means $b > b_0$, the same b_0 for "small" and "large". Thus Figures 6 and 7 (the sociological hypotheses) already describe the model for all b save b_0; Figure 6 applies if $b < b_0$, Figure 7 if $b > b_0$ [p. 331, ZCT]. No evidence or justification is given for this sociological hypothesis, that such a b_0 exists (there are arguments given earlier however to justify a "delay rule" vs. "Maxwell's rule").

In an earlier paper by Zeeman alone on a model for the stock exchange, (paper 11 in ZCT), Zeeman uses the words small and large in the same way as above [p. 364, ZCT]. Thus his hypotheses already give, without using any mathematics at all, the structure of the model as a surface in E^3 save for a 2-dimensional plane described by one control parameter being constant. Nevertheless Zeeman refers to these hypotheses as "disconnected" and "local". He states: "Summarising: we insert seven disconnected elementary local hypotheses into the mathematics, and the mathematics then synthesises them for us and hands us back a global dynamic understanding." [p. 362, ZCT] and elsewhere in that paper, "We now use the deep classification theorem of Thom to synthesise the information acquired so far into a 3-dimensional picture of the surface S . . . " [p. 366, ZCT]. Note that Zeeman is referring to the same theorem that Thom credits to Whitney in "Nature", December 22, 1977. Thom referring to Sussman and Zahler writes: "I would also like to point out a misquotation by the authors. The classification theorem for the "Cusp catastrophe" erroneously quoted as "Thom's theorem" is in this specific case due to H. Whitney."

No justification for the hypotheses of Zeeman's stock exchange model is given in terms of existing data and/or theory of stock exchanges, price theory, etc. In fact no reference to any economic literature is given in this paper.

A defense might still be made that even though Zeeman doesn't justify the stock exchange or war models, still it is good that he has proposed them. Doing so introduces topology (or geometry) into the social sciences and with luck sociologists or economists will be able to develop the models, test them, verify them, etc. Indeed it is important for scientists, social or otherwise, to be aware of mathematical possibilities for models. A positive aspect of all the publicity given to CT is that it may have increased this kind of awareness.

On the other hand good mathematical models are not generated by mathematicians throwing models to sociologists, biologists, etc. for the latter to pick up and develop. Both Thom and Zeeman seem to fit this caricature sometimes in their work or when they give their views on the future of CT in science. Good mathematical models don't start with the mathematics, but with a deep study of certain natural phenomena. Mathematical awareness or even sophistication is useful when working to model economic phenomena for example, but a successful model depends much more on a penetrating study and understanding of the economics.

133

On the other hand around CT, not only does mathematics come first but one sees a sort of mathematical egocentricity; understanding the world is a mathematical (even geometrical) problem. Thom's position on this is clear; e.g. " . . . Eliminate the "obvious" meaning and replace it by the purely abstract geometrical manipulation of forms. The only possible theoretisation is Mathematical." (638, ZCT) or " . . . I agree with P. Antonelli, when he states that theoretical biology should be done in Mathematical Departments; we have to let biologists busy themselves with their very concrete–but almost meaningless–experiments; in developmental Biology, how could they hope to solve a problem they cannot even formulate?" (636, ZCT).

Along with this mathematical egocentricity there is a kind of mystification of the subject that is being created by both Thom and Zeeman. Zeeman does this when he speaks of the "deep classification theorem of Thom" as above and elsewhere in his papers. Presenting this picture to nonmathematicians and even nontopologists has an intimidating effect. Thom does this by using technical mathematical terms without explanation when addressing nonmathematical audiences, and often writing obscurely. Then Zeeman deepens the mystifying power of CT by explaining Thom's obscurities with: "When I get stuck at some point in his writing, and happen to ask him, his replies generally reveal a vast new unsuspected goldmine of ideas" (622, ZCT).

Some defenders of CT may accuse me of discussing very special examples not characteristic of the literature of the subject. I feel that the problem of lack of justification discussed above, is also found in Zeeman's other models. Furthermore Thom's models are even less specific and less developed. On the other hand, Thom's work in CT covers many subjects; in this connection Zeeman writes in his Scientific American article, April 1976, p. 65: "The method has the potential for describing the evolution of form in all aspects of nature, and hence it embodies a theory of great generality."

Corresponding to the universality of CT is a certain superficiality, almost in a dual sense! But one can find even more local maxima in nature than cusps.

It is Thom and Zeeman who have brought CT to the attention of the scientific community with their studies mainly in biology and the social sciences. Thus I can't go along with Ian Stewart's assertion, in "Nature", December 1, 1977, p. 382, "The case in favour of CT rests not on speculative models in the social sciences, but on successful applications to the physical sciences."

I would like to make it clear that I find merit in the Catastrophe theorists use of modern calculus and geometric techniques in models in science. In particular discontinuities can often best be understood via this kind of mathematics. For example it would be important to find a calculus oriented model for the computer, a machine which is intrinsically discrete. Such a calculus model would not be exact, but it could give great insight to automata theory.

There is much value to science in some of the underlying mathematics of CT. I think especially of transversality theory and the theory of singularities of maps. The idea of transversality goes back in history, but took a good development with Pontryagin and Whitney. Thom had used transversality in his early work and by 1956 was putting the concept into a powerful systematic use with his theorem of transversality of jets. In its modern form, *transversality* for a smooth map $f: V \to W$, relative to a submanifold M of W means that whenever $f(x) = y \in M$, $T_x(W) = T_y(M) + Df(x)(T_x(V))$. In other words the derivative of f at x has image complementing the tangent space of M at y. In this case the inverse function theorem implies that $f^{-1}(M)$ is a submanifold of the expected dimension.

Transversality has many ramifications, especially when f itself is a derivative map (or jet map) of some order of another map. These ideas give form to the study of singularities of maps. Again it was Thom who, after fundamental work of Whitney, substantially enriched that theory. Now transversal maps have the property of being an open and dense subset of an appropriate function space. The openness of transversality often can be used to show that corresponding models have the important property of robustness. A model is *robust* if the properties under study remain after perturbation. The approximation properties of transversality imply that one might expect to find transversality present in smooth models. For these reasons it is important for mathematically oriented scientists to be aware of transversality and such examples as cusps. In particular, workers in bifurcation theory from a classical point of view should be (and to some extent are) studying these modern ideas.

Elementary catastrophe theory studies a particular class of maps, (the target space is 1-dimensional, but parameters are allowed) from the above point of view. Sometimes, as in classical mechanics, this situation arises naturally. For example, statics in a simple mechanical system (M, K, V) studies the local minimum of the potential. Here K is kinetic energy or a Riemannian metric on configuration space M while $V: M \to \mathbf{R}$ is the potential energy. Suppose that V depends on a parameter μ in $\mathbf{R}^k$ (i.e., comparative statics). Then ECT is natural for such a study as in ZCT, #17, *A Catastrophe model for the stability of ships*. While this kind of mathematics has a legitimate and even important place in physics, I must object to Zeeman's assertion in the introduction of his ship article (442, ZCT): "it (this example *SS*) is a prototype revealing catastrophe theory as a natural generalization of Hamiltonian dynamics."

More generally the theory of singularities of maps has a constructive role to play in the physical, biological and social sciences. For example comparative statics, as in Paul Samuelson's *Foundations*, studies economic equilibrium prices p satisfying $f(p, \mu) = 0$ where for each parameter value μ, $p \to f(p, \mu)$ is a map from $\mathbf{R}^n$ to $\mathbf{R}^n$. The problem is: how does a solution vary with μ, even for small μ. Since economic theory has shown pretty clearly that the excess demand (in some form) f is not derived from any potential function, CT itself is not relevant. On the other hand, the problem naturally fits into the theory of singularities of maps.

We end this review by a remark on history. Catastrophe theorists often speak as if CT (or Thom's work) was the first important or systematic (CT is "systematic" as a certain study of singularities, but not as a study of discontinuous phenomena) study of discontinuous phenomena via calculus mathematics. My view is quite the contrary and in fact I feel the Hopf bifurcation (1942) for example lies deeper than CT. The Hopf theory shows how a stable equilibrium bifurcates to a stable oscillation in ordinary differential equations. Moreover, there is the reference *Theory of oscillations* by Andronov and Chaiken, 1937, with English translation in 1949 published by the Princeton University Press which is never referred by Thom or Zeeman. This book besides giving an early account of structural stability, gives a good account of dynamical systems in two variables with explicit development of discontinuous phenomena, quite close to Zeeman's use of the cusp catastrophe. Examples from physics and electrical engineering are studied in some depth.

STEPHEN SMALE

ON THE PROBLEM OF REVIVING THE ERGODIC HYPOTHESIS OF BOLTZMANN AND BIRKHOFF

Dedicated to the memory of Rufus Bowen

I would like to develop the idea that by introducing a dissipation/forcing term into Hamilton's equations of physics, one might be able to revive the ergodic hypothesis of Boltzmann and Birkhoff. The compelling fact is that with one main exception (end of section 1), Hamiltonian systems are not ergodic; furthermore other qualitative features of these systems don't seem amenable to study. On the other hand, large classes of general ordinary differential equations have ergodic structurally stable attractors with mixing and other properties. Ergodic theory was developed originally in response to problems arising from Hamilton's ordinary differential equations; but it seems now to have a more natural home in general dynamical systems (at least to the extent it relates to ordinary differential equations).

The goal of this note is to pose the problem of finding a non-Hamiltonian perturbation of a given Hamiltonian system to produce an ϵ-dense ergodic attractor.

Section 1. We give a very brief review of the origins of ergodic theory. But see, e.g., Birkhoff and Koopman in [4], Khinchin [9], or Lanford [10] for a more extended account.

The word *ergodic* was first used by Boltzmann about 100 years ago, at the time he introduced the first version of the "ergodic hypothesis."

Towards the understanding of thermodynamics, Boltzmann was concerned with the statistics of a system of a large number of particles moving according to the laws of mechanics. Thus state space (or phase space) of the system consists of the possible positions and momenta of each particle and the dynamics is described by Hamiltonian ordinary differential equations. Energy is a function on state space and "conservation of energy" asserts that energy is constant under the motion of a state. The flow is thus defined on each surface V_E of constant energy E.

Moreover, there is a natural measure on state space (Liouville) and on each surface of constant energy as well. The flow leaves these measures invariant.

Boltzmann's "ergodic hypothesis" asserted that the trajectory of a single point in phase space passed through every point on that energy surface.

Written for the "International Conference on Non-linear Dynamics" of the New York Academy of Science, Dec. 1979.

137

Before long that was revised to the "quasi-ergodic hypothesis," namely that almost all trajectories were dense on the energy surface.

The quasi-ergodic hypothesis was often accepted under additional conditions on the system; e.g., if the system has no constants other than energy and if the energy surface is bounded.

For example, there is the 1923 paper of Fermi, "Proof that a mechanical normal system in general is quasi-ergodic" [8].

The situation was clarified by Birkhoff's ergodic theorem in 1931 (which in turn was partly inspired by Von Neumann's mean ergodic theorem).

THEOREM (BIRKHOFF ERGODIC THEOREM [4]). *Suppose X is a space with a finite measure and a flow $X \times \mathbb{R} \to X$, $(x, t) \to x_t$, which leaves the measure invariant (e.g., x_t is given as the solution of an ordinary differential equation on X so that $x_t|_{t=0} = x \in X$ is the initial condition). Let $f: X \to \mathbb{R}$ be an integrable function (e.g., $X = V_E$ is an energy surface in phase space and f is a "phase function"). Then except for a set of initial x of measure 0,*

$$\lim_{T \to \infty} \frac{1}{T} \int_0^T f(x_t)\, dt = \bar{f}(x)$$

exists and $\bar{f}: X \to \mathbb{R}$ is integrable and invariant under the flow.

Thus "time averages" exist almost everywhere.

The flow above is called *ergodic* (or metrically transitive in the terminology of Birkhoff) provided every invariant integrable function is constant (almost everywhere).

If the flow on X is ergodic, then according to the ergodic theorem of Birkhoff, $\bar{f}$ is constant and

$$\lim_{T \to \infty} \frac{1}{T} \int_0^T f(x_t)\, dt = \int_{V_E} f$$

is independent of x: time averages equal space averages. Thus an observable (phase function) has a time average independent of initial condition.

The above considerations suggest a new version of the ergodic hypothesis or what Birkhoff called the hypothesis of metric transitivity. We quote from the 1932 article of Birkhoff and Koopman, in [4], p. 465.

> It may be stated in conclusion that the outstanding unsolved problem in the ergodic theory is the question of the truth or falsity of metrical transitivity for general Hamiltonian systems. In other words the *Quasi-Ergodic Hypothesis* has been replaced by its modern version: the *Hypothesis of Metrical Transitivity*.

This hypothesis played an important role in Birkhoff's later work. He not only believed it but part of his work is written assuming that it is true.

Let me quote from his famous "pontifical memoir," "Nouvelles Recherches . . . " in [4], p. 632.

> J'ai réussi plus récemment â demontrer même l'existence des régions annulaires d'instabilité, ce qui indique que l'hypothèse de transitivité est presque certainement remplie dans le cas général. D'ailleurs cette hypothèse a toujours été employée par les physiciens avec grand profit dans la théorie de la mécanique statistique.
>
> C'est pourquoi nous allons nous occuper de la théorie générale d'un tel système transitif . . .

These beliefs held sway in mathematical physics until Kolmogoroff's famous Amsterdam Congress paper in 1954 (see [1]) and subsequent work of Arnold and Moser in 1961–1962 (see [3], [15]). The work of Kolmogoroff, Arnold, and Moser, KAM, showed that near "elliptic" closed orbits of a general Hamiltonian system on an energy surface, ergodicity failed. In that case there exist families of invariant tori of positive measure.

Furthermore these elliptic orbits occur frequently in Hamiltonian systems. Thus the hypothesis of metrical transitivity is false in a definite way (see [1], [13]).

In spite of the work of Kolmogoroff, Arnold, and Moser, there seems to be some kind of ergodicity in physical systems. For example Lebowitz writes [11] p. 41:

> Now I believe that almost all real physical systems are "essentially" ergodic. Indeed this is necessary for understanding why equilibrium statistical mechanics, which includes a description of fluctuations in thermal equilibrium works so well in the real world.

There is one exceptional case of some importance where the system has no elliptic points and ergodicity has been proved. Geometrically, this happens when a particle moves on a manifold of negative curvature (Anosov) [2]. On the physical side is the work of Sinai [19] on ergodicity of a system of elastic spheres. In these cases the ergodicity proofs came out of the framework of the general differentiable dynamical systems of the next section; ideas of E. Hopf were used.

Section 2. Parallel to the growth of interest in Hamiltonian systems, there has been a development of general dynamical systems. Part of the motivation for this development has been from the physical considerations of dissipative and driving forces which destroy the Hamiltonian character as in the Navier–Stokes equations or the Rayleigh dissipation function. Another part of the motivation has to do with the mathematical simplicity and geometry of dealing with unconstrained ordinary differential equations, as the Poincaré–Bendixon theorem. A basic feature of these general dynamical systems (or sometimes simply "dynamical systems," or "differentiable dynamical systems" or the "qualitative theory of ordinary differential equations") is the notion of asymptotic behavior of a (typical) solution. This leads to certain invariant sets and in particular to *attractors* of the dynamical system.

Dynamical systems have two kinds of classical attractors which persist under small perturbations of the differential equations. These are the stable equilibria and the stable nontrivial periodic solutions or oscillators. An important development of recent times is a new kind of attractor which is robust in the sense that its properties persist under perturbations of the differential equation (it is structurally stable).

These new attractors are sometimes called *strange* attractors. The underlying phenomena was seen by Poincaré, Birkhoff, Cartwright and Littlewood, Hadamard, Morse, and others. But more recently a systematic mathematical basis has been laid for their understanding. And complementing that base has been the work of Lorenz [12], May [14], and others on the applied side.

We will give a little of the mathematical side as it applied to our problem here; but see [20], my article in [17], Bowen [5], [6], Ruelle [18], and the cited references for a more complete picture.

Axiom A no-cycle dynamical systems, or for short, Axiom A systems enjoy some satisfying features. First they form a large class and second they are robust and admit analysis. This class includes all the simplest examples which are robust (gradient systems, etc.) and, at the other extreme, those Hamiltonian systems known to be ergodic and mixing. But it also includes a large spectrum of examples in between, dynamical systems with horseshoes, strange attractors, etc. Some important examples are not Axiom A, however, and it is an important problem to expand Axiom A systems to include, e.g., the Lorenz attractor. But the wealth of interesting and useful examples included make Axiom A systems an attractive class.

On the other hand, properties of Axiom A systems persist under perturbation; they are "Ω-stable," i.e., structurally stable on the set of all recurrent solutions. This leads to a number of other good properties. For the definitions and development, the reader may consult the above references. Furthermore, there is the fundamental Bowen–Ruelle theorem, which goes as follows:

First recall that a closed invariant set A is called an *attractor* if all nearby solutions lead to A as $t \to \infty$. The *basin* of A is the union of all solutions which tend to A as $t \to \infty$. An Axiom A system has a finite number of attractors.

THEOREM (BOWEN–RUELLE THEOREM). *For an Axiom A system, except for an initial set of Lebesgue measure* 0, *time averages exist for continuous* *phase functions. More precisely, except for this initial set, a solution $t \to x_t$ tends to some attractor A. The attractor A has a canonical invariant ergodic measure μ and if ϕ is continuous function on the space, then*

$$\lim_{T \to \infty} \frac{1}{T} \int_0^T \phi(x_t)\, dt = \int_A \phi\, d\mu.$$

It is important to note that the theorem applies to almost all solutions with respect to ordinary Lebesgue measure on the space; but the asymptotic behavior is described by a new measure on A.

In the last 25 years, there has been a turnabout in the relationship of ergodic theory to ordinary differential equations. Previously, ergodicity was associated only to constrained ordinary differential equations as Hamiltonian (or volume preserving at least). Nowadays it is not the Hamiltonian, but the general dynamical systems where ergodicity is fitting more naturally.

We would conclude that theoretical physics and statistical mechanics should not be tied to Hamiltonian equations so absolutely as in the past. On physical grounds, it is certainly reasonable to expect physical systems to have (perhaps very small) non-Hamiltonian perturbations due to friction and driving effects from outside energy absorbtion. Today also mathematical grounds suggest that it is reasonable to develop a more non-Hamiltonian approach to some aspects of physics. This leads to the problem we pose in Section 3.

Section 3. Let us formalize the problem of reviving the ergodic hypothesis via the introduction of a dissipative/forcing term. We will do this by posing an abstract idealized mathematical problem, and then discussing some variations and ramifications. In particular the dissipative/forcing term in our statement is any perturbation which balances the energy. We will call an attractor A *ergodic* if it satisfies the conclusion of the Bowen–Ruelle theorem, with the measure positive on open sets of the attractor.

A subset A of X is called ϵ-*dense* if every point of X is within ϵ of some point of A.

MAIN PROBLEM. Given a Hamiltonian system $dx/dt = \xi$ with energy surface V_E and $\epsilon > 0$, is there a new dynamical system $dx/dt = \eta$ on V_E, not necessarily Hamiltonian, ϵ-close to ξ (uniformly with derivatives up to order r), with an ergodic attractor A ϵ-dense in V_E such that almost every point of V_E tends to A as $t \to \infty$?

More broadly, to what extent can such a system $dx/dt = \eta$ be found?

This problem asks for a perturbation of a Hamiltonian system with very strong properties and it might well happen that somewhat less could be true and still be interesting for the goal of reviving the ergodic hypothesis. On the other hand more could be true. The following remarks elucidate these questions.

Remarks: (1) S. Newhouse has pointed out to me that his well-known example can be used to show that a Hamiltonian system cannot always be perturbed to an Axiom A system. Of course if the Hamiltonian system is the geodesic flow of a compact Riemannian manifold with negative curvature, no perturbation is necessary.

One would like a solution of the problem with good robustness properties. And also one would want the attractor to be mixing as well as ergodic.

(2) The results of Newhouse, Ruelle, and Takens [16] are related here. They show that non-Hamiltonian perturbations of uncoupled harmonic oscillators may produce Axiom A strange attractors. Their argument seems to yield more; one obtains an Axiom A system with just one attractor which is ϵ-dense in a single invariant torus of the original Hamiltonian system. One might be able to extend their argument to get the attractor ϵ-dense in the energy surface. But still this would work only for the uncoupled oscillators. The coupled oscillators seem much more difficult as in (3) below. Also for the case of the uncoupled oscillators, a nonstrange attractor could probably be obtained.

(3) A focus for the main problem could be the KAM picture of an area-preserving diffeomorphism of the 2-disk which centers about a general elliptic point (see, e.g., [3], p. 77 or [1], p. 585). Can this diffeomorphism be perturbed to a system with a single ϵ-dense attractor? The same question applies to general 2-dimensional area-preserving diffeomorphisms.

(4) One way of relaxing the problem in a natural way is to not fix the energy surface, but consider perturbations in an ϵ-neighborhood of a given energy surface. In fact there is no reason to demand that the dissipative effort and the driving effect exactly balance to preserve the energy.

(5) The physical side of these matters might constrain the non-Hamiltonian dissipative/forcing perturbation to having a special form. If the original Hamiltonian, for example, came for a simple mechanical system (M, K, V) (M a manifold, K kinetic energy, and V a potential) then especially it would make sense to consider only non-Hamiltonian perturbations with some special structure. Furthermore in this case one might want to study large perturbations.

(6) Another important example is the Navier–Stokes and Euler equations of fluid dynamics (a complicating feature is that state space is infinite dimensional). The Navier–Stokes equations can be regraded as the Hamiltonian Euler equations with an added term which gives a non-Hamiltonian perturbation. Moreover it is often believed that the Navier–Stokes equa-

tions model turbulence. Turbulence according to Ruelle and Takens (see [17] or [18]) corresponds to a strange attractor of the perturbed Hamiltonian system. Furthermore Chorin in his numerical scheme [7] for the Navier–Stokes equations takes a random non-Hamiltonian perturbation of Euler equations and finds statistics that could be explained by the presence of a strange attractor.

While not giving rigorous mathematical backing for our main idea, I find the above considerations supporting the idea of the ergodic hypothesis lying in the framework of non-Hamiltonian perturbations of Hamiltonian systems.

REFERENCES

[1] Abraham, R. and Marsden, J., *Foundations of Mechanics*, 2nd Edition, Benjamin, Reading, 1978.

[2] Anosov, D., Geodesic flows on compact riemannian manifolds of negative curvature, *Proc. Steklov Inst. of Math.* **90** (1967).

[3] Arnold, V. I. and Avez, A. *Theorie Ergodique des Systemes Dynamique*, Gauthier-Villars, Paris, 1967.

[4] Birkhoff, G., *Collected Math. Papers*, Vol. II, Amer. Math. Soc., N.Y., 1950.

[5] Bowen, R., *Equilibrium States and the Ergodic theory of Anosov Diffeomorphisms*, Springer-Verlag, N.Y., 1975.

[6] ———, *On Axiom A Diffeomorphisms*, Amer. Math. Soc., Providence, 1978.

[7] Chorin, A., Numerical study of slightly viscous flow, *J. Fluid Mech.* **57** (1973), 785–796.

[8] Fermi, E., Beweis, dass ein mechanisches Normalsystem in allgemeinen quasi-ergodisch ist, *Phys. Z.* **24** (1973), 261–264.

[9] Khinchin, *Statistical Mechanics*, Dover, N.Y., 1949.

[10] Lanford, O., Erogodic theory and approach to equilibrium for finite and infinite systems, *Acta Physica Austriaca*, Suppl. X. (1973), 619–639.

[11] Lebowitz, J., Hamiltonian flows and rigorous results in non-equilibrium statistical mechanics, in *Proceedings of the 6th IUPAP Conference on Statistical Mechanics*, Rice, Freed, and Light (eds.), University of Chicago, 1972.

[12] Lorenz, E., Deterministic non-periodic flow, *J. Atmos. Sci.* **29** (1963) 130–141.

[13] Markus, L. and Meyer, K., *Generic Hamiltonian Dynamical Systems are neither Integrable nor Ergodic*, Am. Math. Soc., Providence, 1974.

[14] May, R., Simple mathematical models with very complicated dynamics, *Nature* **261** (1976), 459–467.

[15] Moser, J., *Stable and Random Motions in Dynamical Systems*, Princeton Univ. Press, Princeton, 1973.

[16] Newhouse, S., Ruelle, D., and Takens, F., *Occurence of strange Axiom A attractors near quasi-periodic flows on T^m, $m \geqslant 3$*, preprint.

143

[17] Ratiu, T. and Bernard, P. (eds.), *Turbulence Seminar, Berkeley 1976/77*, Springer-Verlag, N.Y., 1977.

[18] Ruelle, D., The Lorenz attractor and the problem of turbulence, in *Turbulence and Navier Stokes Equations*, R. Teman (ed.), Springer-Verlag, N.Y., 1976.

[19] Sinai, Y., Dynamical systems with elastic reflections, *Russ. Math. Surveys* **25** (1970), 137–189.

[20] Smale, S., Differentiable dynamical systems, *Bull. Am. Math. Soc.* **73** (1967), 747–817.

ROBERT EDWARD BOWEN

(February 23, 1947–July 30, 1978)

Rufus Edward Bowen (called Rufus by his friends, because of his striking red hair and beard) died suddenly, of a cerebral hemorrhage, on July 30, 1978. He was not yet thirty-two years of age. During that short lifetime he had already become a mathematician of international stature, and his death shocked the mathematical world.

Bowen was born in Vallejo, and all his schooling was in California. His brilliance was evident quite early, and by the time he received his Ph.D. (at UC, under Smale), he had accumulated a sizable collection of honors and had published half-a-dozen mathematical papers.

The main body of his work was on the ergodic theory of dynamical systems. This has its roots in physics, and Bowen was becoming increasingly interested in mathematical questions which had physical impact.

His work was important. One example of recognition of this was his invited address, at the extraordinarily early age of 27, at the 1974 International Mathematical Congress in Vancouver. He was regarded as one of the leaders of his field.

Rufus was an outstanding and well-appreciated teacher. During his brief career he had 5 Ph.D. students. He was not particularly active in departmental politics, but because of his stature in the mathematical world, he attracted several outstanding visitors in his field to Berkeley, and in this way was beginning to influence the development of the department.

Although rather quiet in politics, as in most things, he nevertheless had clear, independent, and carefully thought out views on many social and political matters. He involved himself actively in the campaign against nuclear weapons.

Rufus was married in 1968 to Carol Twito. They had no children. Their life was simple and unpretentious, punctuated by occasional parties full of noise and dancing. Rufus was a mainstay (if not a star) of the Sunday morning Math Department volleyball games. Although there was a certain amount of travel to other mathematical centers, Rufus mainly liked to stay near home, and his vacations usually took him and his wife to some quiet place in Northern California; it was during one such vacation that his unexpected death came.

His gentle humor, his extraordinary intelligence, his modesty, his utter honesty drew people to him, and he had many friends. They, as well as the University and the mathematical world, will long feel the loss.

Jacob Feldman
Marina Ratner
Stephen Smale

ON HOW I GOT STARTED IN DYNAMICAL SYSTEMS*

1959-1962

1. Let me first give a little mathematical background. This is conveniently divided into two parts. The first is the theory of ordinary differential equations having a finite number of periodic solutions; and the second has to do with the case of infinitely many solutions, or, roughly speaking, with "homoclinic behavior."

For an ordinary differential equation in the plane (or a 2nd-order equation in one variable), generally there are a finite number of periodic solutions. The Poincaré–Bendixson theory yields substantial information. If a solution is bounded and has no equilibria in its limit set, its asymptotic behavior is periodic. The Van der Pol equation without forcing is an outstanding example which has played a large role historically in the qualitative theory of ordinary differential equations.

There was some systemization of this theory by Andronov and Pontryagin in the 1930's when these scientists introduced the notion of structural stability. There is a school now in the Soviet Union, the "Gorki school" which reflects this tradition. Andronov's wife, Andronova-Leontevich, was a member of this school, and I met her in Kiev in 1961 (Andronov had died earlier). A fine book (translated into English) by Andronov, Chaiken, and Witt gives an account of this mathematics.

The work of Andronov–Pontryagin was picked up by Lefschetz after World War II. Lefschetz's book and influence helped establish the study of structural stability in America.

The second part of the background mathematics has to do with the phenomenon of an infinite number of periodic solutions (persistent under perturbation) or the closely related notion of homoclinic solutions. Again, Poincaré wrote on these problems; Birkhoff found a deeper connection between the two concepts, homoclinic solutions and an infinite number of periodic points for transformations of the plane. From a different direction, Cartwright and Littlewood, in their extensive studies of the Van der Pol equation with forcing term, came across the same phenomena. Then Levinson simplifed some of this work.

2. Next, I would like to give a little personal background. I finished my thesis in topology with Raoul Bott in 1956 at the University of Michigan; and that summer I attended my first mathematics conference, in Mexico City, with my wife, Clara. It was an international conference in topology and my introduction to the international mathematics community. There I

---

*Partly based on a talk given at a Berkeley seminar circa 1976.

met René Thom and two graduate students from the University of Chicago, Moe Hirsch and Elon Lima. Thom was to visit the University of Chicago that fall, and I was starting my first teaching job at Chicago then (not in the mathematics department, but in the College). In the fall I became good friends with all three. I attended Thom's lectures on transversality theory and was happy that Thom and Hirsch became interested in my work on immersion theory.

My main interests were in topology throughout these years, but already my thesis contained a section on ordinary differential equations. Also at Chicago, I used an ordinary differential equation argument to find the homotopy structure of the space of diffeomorphisms of the 2-sphere. Having both Dick Palais and Shlomo Sternberg at Chicago was very helpful in getting some understanding of dynamical systems.

3. It was around 1958 that I first met Mauricio Peixoto. We were introduced by Lima who was finishing his Ph.D. at that time with Ed Spanier. Through Lefschetz, Peixoto had become interested in structural stability and he showed me his own results on structural stability on the disk D^2 (in a paper which was to appear in the *Annals of Mathematics*, 1959). I was immediately enthusiastic, not only about what he was doing, but with the possibility that, using my topology background, I could extend his work to n dimensions. I was extremely naive about ordinary differential equations at that time and was also extremely presumptuous. Peixoto told me that he had met Pontryagin, who said that he didn't believe in structural stability in dimensions greater than two, but that only increased the challenge.

In fact, I did make one contribution at that time. Peixoto had used the condition on D^2 of Andronov and Pontryagin that "no solution joins saddles." This was a necessary condition for structural stability. Having learned about transversality from Thom, I suggested the generalization for higher dimensions: the stable and unstable manifolds of the equilibria (and now also of the nontrivial periodic solutions) intersect transversally. In fact, this was a useful condition, and I wrote a paper on Morse inequalities for a class of dynamical systems incorporating it.

However, my overenthusiasm led me to suggest in the paper that these systems were almost all (an open dense set) of ordinary differential equations! If I had been at all familiar with the literature (Poincaré, Birkhoff, Cartwright–Littlewood), I would have seen how crazy this idea was.

On the other hand, these systems, though sharply limited, would find a place in the literature, and were christened Morse–Smale dynamical systems by Thom. This work gave me entry into the mathematical world of ordinary differential equations. In this way, I met Lefschetz and gave a lecture at a conference on this subject in Mexico City in the summer of 1959.

We had moved from Chicago in the summer of 1958 to the Institute for Advanced Study in Princeton; Peixoto and Lima invited me to Rio to finish the second year of my NSF postdoctoral fellowship. Thus, Clara and I and our two kids, Nat and Laura, left Princeton in December, 1959, for Rio.

4. With kids aged $\frac{1}{2}$ and $2\frac{1}{2}$, and most of our luggage consisting of diapers, Clara and I stopped to see the Panamanian jungles (from a taxi). I had heard about and always wanted to see the famous railway from Quito in the high Andes to the jungle port of Guayaquil in Ecuador. So, around Christmas of 1959, the four of us were on that train. Then we spent a few days in Lima, Peru, during which we were all quite sick. We finally took the plane to Rio. I still remember well arriving at night and going out several times trying to get milk for the crying kids, always returning with cream or yogurt, etc. We learned later that milk was sold only in the morning in Rio.

But with the help of the Limas, the Peixotos, and the Nachbins, life got straightened out for us. In fact, it happened that just before we arrived the leader of an abortive coup, an air force officer, had escaped to Argentina. We got his luxurious Copacabana apartment and maids as well.

Shortly after arriving in Rio my paper on dynamical systems appeared, and Levinson wrote me that one couldn't expect my systems to occur so generally. His own paper (which, in turn, had been inspired by work of Cartwright and Littlewood) already contained a counter-example. There were an infinite number of periodic solutions and they could not be perturbed away.

Still partly with disbelief, I spent a lot of time studying his paper, eventually becoming convinced. In fact this led to my second result in dynamical systems, the horseshoe, which was an abstract geometrization of what Levinson and Cartwright–Littlewood had found more analytically before. Moreover the horseshoe could be analysed completely qualitatively and shown to be structurally stable.

For the record, the picture I abstracted from Levinson looked like this:

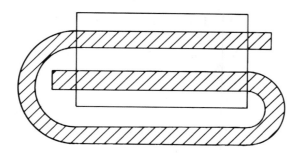

When I spoke on the subject that summer (1960) at Berkeley, Lee Neuwirth said: "Why don't you make it look like this?"

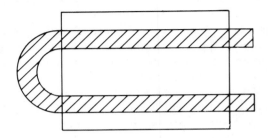

I said "fine" and called it the horseshoe.

I still considered myself mainly a topologist, and when considering some questions of gradient dynamical systems, I could see possibilities in topology. This developed into the "higher-dimensional Poincaré conjecture" and was the genesis of my being quoted later as saying I did my best-known work on the beaches of Rio. In fact, I often spent the mornings on those beaches with a pad of paper and a pen. Sometimes Elon Lima was with me. In June I flew to Bonn and Zurich to speak of my results in topology. This turned out to be a rather traumatic trip, but that is another story.

I had accepted a job at Berkeley (at about the same time as Chern, Hirsch, and Spanier, all from Chicago) and arrived there from Rio in July, 1960. Except for a few lectures on "the horseshoe," I was preoccuped the next year with topology. But, in the summer of 1961, I announced to my friends that I had become so enthusiastic about dynamical systems that I was giving up topology. The explicit reason I gave was that no problem in topology was as important and exciting as the topological conjugacy problem for diffeomorphisms, already on the 2-sphere. This conjugacy problem represented the essence of dynamical systems, I felt.

5. During this year I had an irresistible offer from Columbia University, so we sold a house we had just bought and moved to New York in the summer of 1961. But before taking up teaching duties at Columbia, I spoke at a conference on ordinary differential equations in Colorado Springs and then in September, 1961, went to the Soviet Union. At a meeting on nonlinear oscillations in Kiev, I gave a lecture on the horseshoe example, "the first structurally stable dynamical system with an infinite number of periodic solutions." I had a distinguished translator, the topologist, Postnikov, whom I had just met in Moscow. Postnikov agreed to come to Kiev and translate my talk in return for my playing go with him. He said he was the only go player in the Soviet Union. My roommate in Kiev was Larry Markus.

I met and saw much of Anosov in Kiev. Anosov had followed the Gorki school, but he was based in Moscow. After Kiev I went back to Moscow where Anosov introduced me to Arnold, Novikov, and Sinai. I must say I was extraordinarily impressed to meet such a powerful group of four young mathematicians. In the following years, I often said there was nothing like that in the West.

I gave some lectures at the Steklov Institute and made some conjectures on the structural stability of certain toral diffeomorphisms and geodesic flows of negative curvature.

After I had worked out the horseshoe, Thom brought to my attention the toral diffeomorphisms as an example with an infinite number of periodic points which couldn't be perturbed away. Then I had examined the stable manifold structure of these dynamical systems.

6. After teaching dynamical systems the fall semester at Columbia, I was off again, with Clara and the kids, this time to visit André Haefliger in Lausanne for the spring quarter. Besides lecturing in Lausanne, I gave lectures at the Collège de France, Urbino, Copenhagen, and, finally, Stockholm, all on dynamical systems and emphasizing the global stable manifolds, which I found more and more to lie close to the heart of the subject. In Stockholm, at the International Congress, I saw Sinai again and he told me that Anosov had proved all the conjectures I had made the preceding year in the Soviet Union.

In Lausanne I had begun to start thinking about the calculus of variations and infinite-dimensional manifolds, and this preoccupation took me away from dynamical systems for the next three years.